Springer Series in Synergetics   Editor: Hermann Haken

Synergetics, an interdisciplinary field of research, is concerned with the cooperation of individual parts of a system that produces macroscopic spatial, temporal or functional structures. It deals with deterministic as well as stochastic processes.

Self-Organization and Management of Social Systems

Insights, Promises, Doubts, and Questions

Editors:
H. Ulrich and G. J. B. Probst

With 30 Figures

Springer-Verlag
Berlin Heidelberg New York Tokyo 1984

Professor Dr. Hans Ulrich
Dr. Gilbert J. B. Probst

Institut für Betriebswirtschaft, Hochschule für Wirtschafts- und Sozialwissenschaften
CH-9000 St. Gallen, Switzerland

Series Editor:
Professor Dr. Dr. h. c. Hermann Haken
Institut für Theoretische Physik der Universität Stuttgart, Pfaffenwaldring 57/IV,
D-7000 Stuttgart 80, Fed. Rep. of Germany

ISBN-13: 978-3-642-69764-7   e-ISBN-13: 978-3-642-69762-3
DOI: 10.1007/978-3-642-69762-3

2153/3130-543210

Foreword

Self-organization of systems belonging to quite different disciplines has been a central topic of synergetics since its beginning. I am therefore particularly pleased that Hans Ulrich and Gilbert Probst have not only undertaken to organize an interdisciplinary meeting on Self-Organization and Management of Social Systems, but have also edited these articles written by leading scientists after and based upon that symposium.

While the previous volumes of the Springer Series in Synergetics were mainly devoted to physical, chemical and biological systems, with only the book by W. Weidlich and G. Haag dealing with "Quantitative Sociology" (Springer Ser. Syn., Vol. 14), the present volume opens a new perspective.

As the reader will notice, the multitude of facets of self-organization is well reflected by various authors belonging to different disciplines and representing different schools of thought. When such a wide scope of fields - ranging from physics to sociology - is covered, it is not surprising that the existence of a "hiatus" between sociology and the natural sciences was felt by some participants. But in contrast to their opinion, I do not believe that this is caused by a different complexity of the systems. The brain of higher animals (not to mention the human brain) is, in my opinion, at least as complex as a social system, but it can be considered as an object of study by the natural sciences. I rather believe that the "hiatus" is caused by the fact alluded to on various occasions at this meeting, that in sociology the researcher becomes part of the system. Or, more generally speaking, that we are inclined to make a fundamental difference between "human" and other sytems from the very beginning. Certainly there are ethical, moral and other aspects of social systems which are not shared by any other system. Yet it is my deep conviction that at a sufficiently high level of abstraction, sociological processes can be subsumed under principles of general validity in the animate and inanimate world, not because of superficial analogies between, say, a physical system and a sociological system, but because of deep-rooted structural laws.

But I do not want to induce any bias in the reader - who is himself a self-organizing system. I rather hope that he will be fascinated - as the participants of this symposium were - by the many intriguing aspects of this interdisciplinary endeavour, and I wish to congratulate the editors of this volume, which represents an important contribution to the discussion between sociology and the natural sciences.

Hermann Haken

Preface

This book is about mechanisms of control in its broadest sense, especially the ability of systems to maintain identity and autonomy and to keep their relevant variables within "physiological" limits. The authors try to answer questions such as what orderedness of a system is based on, how this orderedness has arisen, how it can be maintained and altered and further developed. Self-organization is a main phenomenon in systems maintaining their identity and autonomy. Research results and epistemological consequences of these results are presented and discussed. Heinz von Foerster, a pioneer of cybernetic research and bio-engineering, Francisco Varela, a biologist and founder of autopoiesis and self-referential processes, Hermann Haken, physicist, the "father" of laser theory and synergetics, Rupert Riedl, zoologist and marine biologist and one of the main representatives of an evolutionary epistemology, and Peter Hejl (sociologist), Hans Ulrich, Fredmund Malik, Gilbert Probst (management scientists) and Peter Dachler (psychologist), all give interesting and new insights and ask nagging questions concerning principles of self-organization.

But the book also is about management of social systems and the importance and consequence a phenomenon such as self-organization has for management. In many of the current textbooks, management is defined as a list of individually performed functions, like planning, decision making, organizing, leadership and control. The point of departure of such a view is usually the individual manager, and the central focus is to investigate how the manager plans or takes decisions and how he should perform these functions for greater organizational effectiveness. We believe, however, that such an analytical and individualistic approach is unsuitable for an understanding of the true character and meaning of management, particularly with respect to managing complex social systems. The particular activities of a manager become meaningful only if they are studied in a broader context. These individually performed management functions have to be seen as activities whose meaning and impact can only be understood within the comprehensive social system of which the performing manager is a part. These systems which we define as purposeful social systems are creations of human society. Members of this class of purposeful social systems are not just business enterprises, but also schools, hospitals, government organizations, churches, various social associations, etc. In our context we can understand these institutions as systems which do not come into existence naturally, that is without human intentions and actions, but which are products of human civilization. Although a part of society, they have to be able to act as an autonomous entity. However, the ability for the system to act presupposes that it is the individual members who act in determining the system's form, processes and outcomes, that is the system's actions.

Although social systems do not come into existence naturally, they are often the result of human action but not of human design or intent (see Friedrich von Hayek). Self-organization is a phenomenon that can be observed in physical, biological and in social systems.

The various contributions contained in this volume were written after an interdisciplinary research colloquium which took place at the University of St. Gall, Switzerland, called 1st "St. Galler Forschungsgespräche" on Management and Self-

Organization in Social Systems (14-16th September 1983). The authors have attempted not only to state their pre-colloquium positions but also to incorporate insights gained during the intensive discussions that characterized the colloquium. A few basic convictions were shared by all participants, irrespective of their research interests: the insight that interdisciplinary collaboration should not only be called for but also be realized through one's own efforts; the conviction that the phenomenon of self-organization is of central importance for many areas of knowledge; and the common interest in questions of epistemology and research strategy which result from recently acquired knowledge in many disciplines. If such an unusual discussion across the boundaries of individual sciences was possible, then this is mainly due to the willingness of the authors of this book, despite their considerable geographical distance from St. Gall and manifold other obligations, to take part in an intensive colloquium in a small group, and, in addition, to find the time to put their ideas down in writing. Our thanks go also to the Rectorate of the St. Gall Graduate School, especially to Prorector J. Anderegg, whose initiative and energy were central to the creation of the kind of framework without which this type of interdisciplinary exchange of ideas is impossible.

St. Gall, July 1984   *Hans Ulrich* and *Gilbert J.B. Probst*

Contents

Part I

Principles of Self-Organization in Physical, Biological, and Social Systems

Principles of Self-Organization – In a Socio-Managerial Context

H. von Foerster

One Eden West Road, Pescadero, CA 94060, USA

0. Opening

I have to confess that when I first received the kind invitation from Dr. Probst to participate in a meeting entitled "Management and Self-Organization in Social Systems" I was not quite clear about my role in such a meeting. I am not a stranger to the notion of Self-Organization; but when I considered it in the context of management and, moreover, in the environment of a Hochschule für Wirtschafts- und Sozialwissenschaften, I felt lost. I understand so little about management that already in grade school my teachers complained that this boy is unmanageable. In fact, I had to look "management" up in my dictionary (1). Here I found that it is derived from ... "constraining the movement of hands", having the same root as "to manacle", that is, putting someone into handcuffs : I was prepared to decline this invitation.

Fortunately not much later the organizers of this meeting sent me a paper by Messrs. Malik and Probst entitled "Evolutionary Management" (2), apparently with the idea of giving me a clue of what this meeting would be about. There are two mottos that initiate this paper. Since after I read them I knew I would accept the invitation, I shall read them also to you. The first is a quote by Peter Drucker who, like me, grew up in Vienna, and whose parents happened to be good friends with mine:

"The only things that evolve by themselves in an organization are disorder, friction, and malperformance ..."

That is not a bad start for a paper that addresses itself to self-organization in management. The second motto is again by a Viennese, the Nobel laureate Friedrich von Hayek, who participated in a conference on principles of our topic I had organized almost a quarter century ago. Here is his quote:

"... the only possibility of transcending the capacity of individual minds is to rely on those super-personal 'self-organizing' forces which create spontaneous order".

With these two mutually anihilating quotations the organizers of this meeting had me almost hooked, but succeeded completely after I had read the entire article. There were four points that were very much to my liking:

  (i)   Hierarchies are inappropriate skeletons for a managerical structure;
  (ii)   The importance of flexibility and adaptation;
  (iii) Limited control of, and knowledge in, the system;
  (iiii) And finally, the last line of this article which reads:

"As managers we have to ... learn to be what we really are: not doers and commanders, but catalysts and cultivators of a self-organizing system in an evolving context."

I found myself very close to this sentiment, an affinity with a point I once made at the end of one of my papers (3). I called it an "ethical imperative":

"Act always so as to increase the number of choices!"

My general impression was that the two authors were in search of an epistemology; an epistemology that takes account of the situation in which the manager is himself an element of the system he is managing.

A decade or two ago nobody in his right state of mind would have dared to consider this problem, or even to formulate it that way. And if one would have done so, all experts would have had the times of their life to show that this self-inclusion is the root of all paradox. If mildmannered they would have referred to the barber in the village who shaves all who do not shave themselves (clearly, those who shave themselves need not to be shaved). So far so good. But should the barber shave himself? Of course not, for he shaves only those who do not shave themselves. Apparently, he is not to shave himself. But then ... etc. If it is a learned expert he may cite Bertrand Russel's victory over the paradoxical "set of all sets that do not contain themselves as elements" (with the unanswerable question: does this set contain itself as element, or does it not?). This victory was celebrated as the "theory of types" in which this liberal gentleman simply forbade self-inclusion on logical grounds (a proposition must be either true or else false; here, however, these propositions are true when apprehended as false, and false when apprehended as true).

Fortunately, today the situation is quite different, thanks to the pioneering work of three gentleman. One is Gotthard Günther, a philosopher, now professor at the University of Hamburg, who developed a most fascinating multi-valued logical system (4), quite different from those of Tarsky, Quine, Turquette, and others. Then, there is Lars Löfgren, a logician in Lund, Sweden, who introduced the notion of "autology" (5), that is, concepts that can be applied to themselves, and in some cases need themselves to come into being. I shall dwell on these points in a moment. Finally, we have the work by Francisco Varela, who sits right here, who, as you all know, expanded G. Spencer-Brown's Calculus of Indication to become a Calculus of Self-Indication (6).

My plan for this paper is to build upon these ideas, and in attempt to maximize my usefulness to this meeting, I shall present my points complementary to those made by Malik and Probst in their paper (reprinted in this volume):

(i) First, I shall expand on the notion of autology;

(ii) Second, I shall give a brief account of a rather general interpretation of the concept of computation, and its (conceptual) realization in form of "machines", because I need this concept for the next point I wish to make, namely,

(iii) Recursive Computations.

(iiii) Finally, I shall make use of all that by talking about self-organization in the socio-managerial context.

1. Autology

I wish to contemplate the manager who considers himself a member of the organization he manages. If he takes this consideration seriously, he has to apply his managerial perceptions and acts to himself, to his own perceptions and acts. Management, clearly, is an autological concept. In some other context, such concepts are referred to as "second-order concepts".

3

To get a feeling for the peculiar logical properties that distinguish autologies from other concepts, I invite you to participate in the experiment suggested in Figure 1. Kindly follow the instructions as given in the caption of this Figure, and do not give up until indeed the black spot has completely disappeared. This phenomenon is usually referred to as the "blind spot" in our visual field, and physiologists have a straightforward explanation for this phenomenon (Figure 2). There is a place on our retina where there are no receptor cells, neither rods nor cones. This place is called the "disc", and it is there where the optic nerve leaves the eye ball. Of course, the black spot can not be seen when one is forced to project the spot on the disk when keeping the asterisk focused on the fovea.

Figure 1: Hold paper with right hand. Close left eye. Fixate asterisk. Move paper to and fro along the line of vision. Watch black spot disappear (at eye-paper distance between 12 and 14 inches). Keep asterisk fixated and move paper slowly parallel to itself up, down, left, right, or in circles: black spot remains invisible.

This explanation seems to take care of these affairs, and we could turn to other matters. However, I would like to make two comments here, one regarding the blind spot phenomenon itself, the other about this explanation.

What apparently is surprising in this experiment is its demonstration of the incompleteness of our visual field, an imcompleteness of which we are totally unaware under normal conditions. If one were to stress now the autological nature of visual perception or, as a matter of fact, of perception in general, one may say that we don't see that we don't see!

This suggests that the problem here is not not-seeing, the problem is not seeing that one is not seeing. This is a problem of the second order, and it is graciously overlooked in the orthodox explanation above. Hence, not seeing the problem is the blind spot phenomenon all over again, only now on the cognitive level.

My strategy of introducing second-order concepts containing negatives was to show at once their unusual logical structure, for here double negation does not yield affirmation: not not-seeing does not imply seeing.

I shall now turn to examples of these concepts with an affirmative logical skeleton, again to draw your attention to the different "logical types" as Gregory Bateson may have said, of notions that are embedded in their own domain.

Let me begin with "purpose". If taken as a first-order concept one may speak of something "having a purpose". However, taken on its second-order level we may ask "what is the purpose of 'purpose'?", that is, to ask why introduce the notion of purpose in the first place. Of course, the answer here is straightforward, namely, to avoid contemplating variable and unpredictable trajectories by attending to a more or less invariant state of affairs; the "goal", the "end", *telos*. However, by paying attention to the autological nature of "purpose", our gaze is shifted from "something", the observed, to "somebody", i.e., the one who uses this term, that is, the observer (7).

Next, I turn to language: "What is language? Or better, what is "language"? Whatever is asked here, it is language we need for an answer; and, of course, we need language to ask that question on language. Hence, if we did not know the answer,

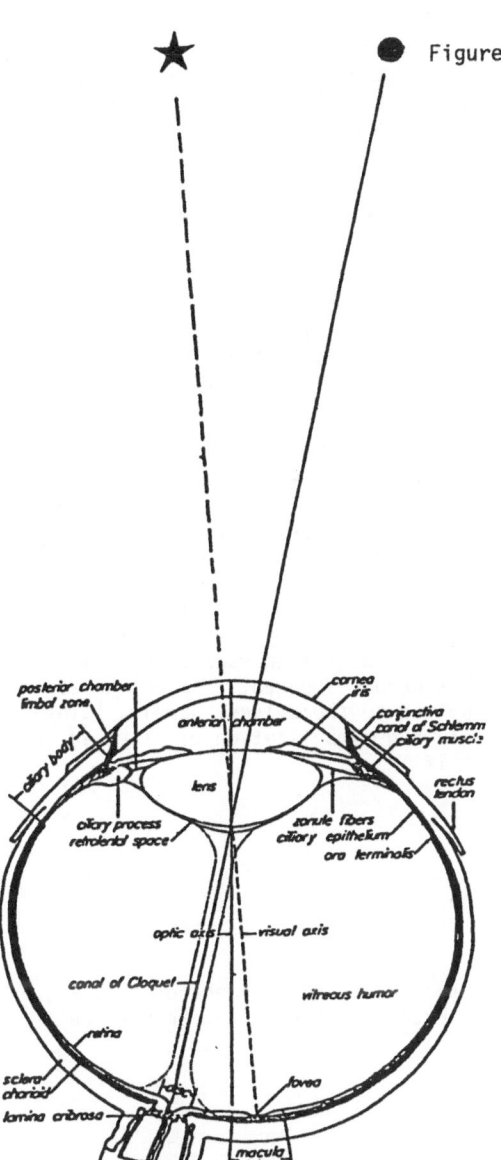

Figure 2: Horizontal section of the right human eye, showing locus of projections.

how could we have asked the question in the first place? and if indeed we did not know it, what will an answer be like that answers itself?" (8). The semantic loop I am stressing here suggests a logical constraint in a possible definition of "language", namely, its autological nature. That is, for any referential communicative conduct to be "language", it must contain reference to its communicative conduct (i.e., a language must be able to express the notion of "language" or, as Humberto Maturana is fond of saying, language must be able to refer to its referring, must able "to point to the pointing"). Of course, the ultimate teaser in this context is Ludwig Wittgenstein's question (9): "What is a question?", and I will leave it to you to tackle it.

As a final example, I shall now deal with the autological nature of the central topic of our meeting, namely "organization". Let me again go through the shift from a first-order to a second-order interpretation of this concept. We take the corresponding transitive verb "to organize", then we stipulate a world in which the organizer and his organization are as fundamentally separated from one another as are the active and the passive forms; it is the world of organizing the other, it is the world of the injunction:

"Thou shalt ..."

On the other hand, if we contemplate the organization of an organization so that the one slips into the other, i.e., "self-organization", we stipulate a world where the actor acts ultimately on himself, for he is included in his organization: it is the world of organizing oneself, it is the world of the injunction:

"I shall ..."

From this it appears to be clear that shifting from first to second-order interpretations has as one of its consequences a shift in the epistemological foundations of ethics. The novelty appears in the latter case, where for the first time one may begin to see the ethical epistemologist becoming accountable for his own epistemology.

I hope that with all these examples of autology, and most explicitly in that of self-organization, my position not to yield to the Russellian escape route into meta-domains (e.g., "meta-languages", etc.) has become apparent. May be the essential feature of those concepts that can be applied to themselves, namely, "closure", has become apparent as well. Perhaps the following symbolization of, say, an organization that applies its competence to itself

ORG

suggest "closure" even more persuasively.

Moreover, those of you who are familiar with the formal development of this argument may recognize in the "recursive pointer" Francisco Varela's mark for the autonomous state

which he introduced almost ten years ago in his seminal paper on a calculus of self-reference (6).

While at first one would think that the introduction of closure adds richness to the arguments, it does in fact do the opposite. It removes one degree of freedom. This is so, for whatever we may consider the "end" in any domain, it must coincide with the "beginning", otherwise the system is not closed. Since this is a crucial point, as you will see in a moment, let me demonstrate this on two examples.

The first I will take from physics, from the early days of wave mechanics. As you may remember, some experiments with elementary particles, electrons in particular, suggested that they could be interpreted as the particles behaving like waves, augmenting each other when crests meet crests and valleys meet valleys; but annihilating each other when crests meet valleys. If this is so, de Broglie argued, electrons orbiting the nucleus in an atom would always annihilate themselves, unless they would move in orbits that are integer multiples of their wavelength (see Figure 3), only then crests would meet crests, and valleys valleys; that is, the end of a wave train must be its beginning.

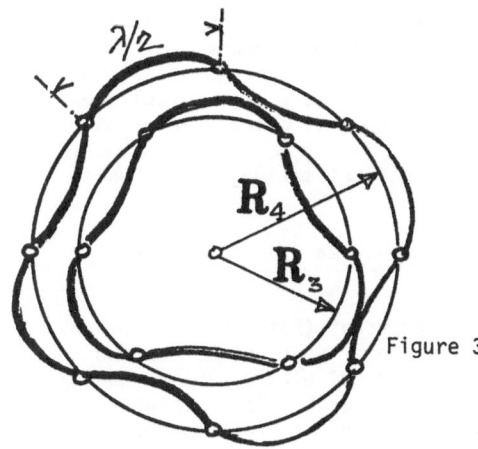

Figure 3: Stable electron orbits along "Eigen-Radii" corresponding to circumferences of multiples of wavelength λ:

$$R_3 = 3\lambda/2\pi; \quad R_4 = 4\lambda/2\pi.$$

With this condition to be fulfilled, it is clear that only certain orbits can exist, they are "quantum jumps" apart, and it was the confirmation of de Broglie's hypothesis through quantum physics that brought him the Nobel Prize.

Please note again from the argument or from Figure 3 that the condition of closure, i.e. the end fitting the beginning, carves out from the infinite possibilities these electrons could move around their nuclei, a set of discrete solutions whose values fulfill the desired condition.

These values are called "Eigen-values" ("self-values"), first so called around the turn of the century by the mathematician David Hilbert in connection with solutions of problems with similar logical structure.

My second example has to do with self-referential propositions. As you may remember, these have always been believed to be the real trouble makers, for instance, the paradoxes of the Epimenides type, one of which I have mentioned before (the barber's difficulty of shaving himself). However, as we shall see in a moment, these situations are not only not irresolvable, as was thought before, but their solutions provide us with insights into other domains.

Consider the following incomplete sentence:

    THIS SENTENCE HAS ... LETTERS

and find a number whose name spelled out and inserted into the blank spaces makes this sentence complete and consistent. Clearly, from the infinite reservoir of numbers only a few, if any at all, will fulfill this condition. For instance THIRTY would not do, for the sentence "This sentence has thirty letters" has in fact only 28 letters.

There are two solutions, two "Eigen-values", to this problem which satisfy the conditions of above. One of these is THIRTYONE. Indeed, the sentence

THIS SENTENCE HAS THIRTYONE LETTERS

has 31 letters. Moreover, note that this sentence says what it does!

The other solution I suggest you work out for yourself, because such an exercise drives home forcefully what it means "to make ends meet" (10).

Since in these cases of closure one runs the result of an operation again through this operation, one speaks of "recursive operations" (from re=again, and currere= to run). The theory that provides the formalism for these processes is called "recursive function theory". Today, this mathematical field is a well established and extensive body of knowledge (11), and I shall touch upon it briefly later on.

What are the consequences of all this for management? Let me suggest one which I think has many ramifications:

In a self-organizing managerial system each participant is also a manager of this system.

Such an organizational structure is called a "heterarchy" (heteros = the other, and archein = to rule), for at one time it may be one of your neighbours who is making the decisions, at another you, as the neighbour of others. This organization is, of course, the antipode of a "hierarchy", where the "holy" (hieros) rules, where the boss has all the power, and the line of command is from top down.

The notion of heterarchy was, to my knowledge, first introduced by Warren McCulloch in one of his papers "A Heterarchy of Values Determined by the Topology of Nervous Nets" (12), which to read is an intellectual feast.

As McCulloch told it, he derived the concept of a heterarchy from a principle he very much cherished. It is

The principle of potential command, where information constitutes authority.

As an example of this principle he used to tell the story of the battle of the Midway Islands where the Japanese fleet was out to destroy the American fleet. Indeed, the American flagship went down in the first few minutes, and its fleet was left to organize itself, i.e., to switch from a hierarchy to a heterarchy. What happened now was that the skipper of each vessel, small or big, took the command over the entire fleet whenever he realized that he, because of his position at that moment, knew best what to do. The result, as we all know, was the destruction of the Japanese fleet, and the turning point for the developments of the war in the Pacific.

2. Machines

I am sure you spotted in my presentation the two main themes to which I returned again and again, self-reference and closure, and also sensed my attempt of slipping these two notions into each other. The device I employed in this attempt was "recursion", and I hope that you could taste some of its flavour, because I would like now to demonstrate the power of this concept in the context of our meeting. Since I wish to do that by invoking elementary steps in its formalism, the formalism of recursive computations, I will make first some preliminary remarks on computation in general.

First, let me remind you that the etymological root of "computation" does not in the least confine it to numerical expressions. The word is a merger of "com" = together, and of "putare" = to contemplate, that is, contemplating things together. Clearly, there is no restriction regarding the "things" contemplated, and I shall use it in this general sense.

As a vehicle for talking about computation I am going to use the idea of a "machine", very much in the sense in which Alan Turing introduced it almost a half century ago, namely, as a conceptual device with well-defined rules of operation. However, I will not describe here a Turing Machine (13), for it would move us too far away from our central topic, but I will give you an account of even more general conceptual computing devices, the so-called "Finite State Machines" (14).

Of those there are two kinds available now, the Trivial and the Non-Trivial Finite State Machine, or the TM and NTM for short. I shall first extoll the charms of the trivial machine (TM), and then develop those of the NTM.

The Trivial Machine

Figure 4 is a schematic representation of a TM, with the labels x, y, f, referring to "input", "output", and "function" of this machine respectively, and the arrows indicating the direction in which the operations are performed. The idea is to have a clear understanding of process. Take, for example, x and y representing the natural numbers 1, 2, 3, 4, ..., and let the function of this machine be the production of an output y that is the square of the input x, i.e., this machine is a "Squaring" TM. Of course, you know what is going on here, and you also know that there is a variety of ways for describing this, some anthropomorphic - or even biomorphic - ways. For instance, if one "feeds" our Squaring machine a 4 (x=4), it will "spit out" 16 (y=16). Or take another TM, those one sees today at the checkout counters of supermarkets. An item is moved with its code lines over the machine's "sensor" and the printer enters "NOODLES ... $ 3.50" on the bill (a "Billing" TM). Or kick a ball into the air (x=kick) and watch it flying up and falling down (watch y). This is the operation of a "Gravitational-Attraction" TM. Or consider the structure of the deductive syllogism. The classical example is, of course: "All men are mortal" (the major premise); "Socrates is a man" (the minor premise); and how the conclusion: "Socrates is mortal", I call this the "All-Men-Are-Mortal" trivial machine, for whatever you take as an input, as long as it is a man, a (potential) corpse will emerge on the other side; and so on and so forth.

Figure 4: Trivial Machine .

I have chosen this outrageous mix of samples, for I wanted to let the following three points to become utterly, utterly clear.

Number one: In spite of the tremendous variety of context in these examples, the underlying schema of argument, logic, operation, etc., is in all the same: because of the invariable relationship (f) between input (x) and output (y), a y once observed for a given x will be the same for the same x given later. The consequence of this is that all TMs are:

(i) predictable,
(ii) history independent.

Number two: Because of the popularity of the inference schema of trivial machines the three the machine determining entities, x, y, and f, depending on the different contexts, appear and re-appear under the most diverse names. Here is an incomplete list:

```
         x            f            y
       input      operation     output
independent variable  function  dependent variable
      cause      Law of Nature    effect
  minor premise  major premise      conclusion
      stimulus      C.N.S.      response
    motivation    character     deeds
        goal        system      action
       ...           ...          ...
                     ...          ...
```

Number three: When a TM is synthesized, that is, when the x - y correspondence
(i.e., the function f) is established, this machine is then unambiguously defined.
One speaks here of a synthetically determined system. A particularly nice feature
of these machines is that they are also analytically determinable, for one simply
has to record for each given x the corresponding y. This record is then "the machine".
Hence, all TMs are

(iii)   synthetically deterministic,
(iiii) analytically determinable.

I shall summarize this now by inviting you to contemplate a trivial machine that
has the following properties: it can distinguish four input states (x): A, U, S, T;
and two output states (y): 0, 1. The correspondence between x and y is established
through this Table:

```
              f

           x │ y
           ──┼──
           A │ 0
           U │ 1
           S │ 1
           T │ 0
```

Hence, from the input sequence of, say, A, U, S, T, the machine will compute the
output sequence 0, 1, 1, 0; or from the sequence U, S, A, it will compute 1, 1, 0;
and when this sequence is repeated again and again, undisturbed of what may happen
in between, we shall obtain again and again, 1, 1, 0, until the Day of Judgement.

Non-Trivial Machines

Obedience is the hallmark of the trivial machine; it seems that disobedience is
that of the non-trivial machine. However, as we shall see, the NTM too is obedient,
but to a different voice. Perhaps, one could say obedient to its inner voice.

How do NTMs differ from TMs? In fact in a very simple, but profoundly consequential
way: a response once observed for a given stimulus may *not* be the same for the same
stimulus given later.

The most fruitful way to account for such changes in performance may be through
the machine's internal states (z), whose values co-determine its input-output relation (x, y). Moreover, the relationship between the present and subsequent internal states (z,z') is co-determined by the inputs (x). Perhaps the best way to visualize this is by seeing this arrangement as a machine in a machine (see Figure 5).
From the outside such a machine looks very much like a trivial machine, with an input x and an output y. However, when the lid is taken off (as in Fig. 5), one can
see now the entrails of an NTM. The novel feature here is the place (circle in the
centre) that holds the internal state z. This state, together with the input x,

10

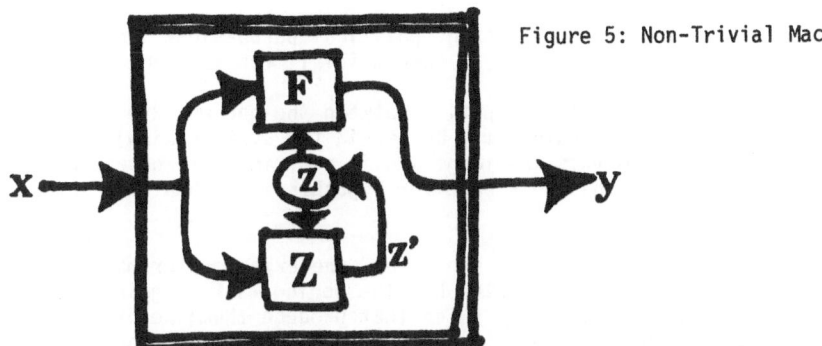

Figure 5: Non-Trivial Machine.

furnish an input - on the one hand - to **F**, a trivial machine computing the NTM's output y, and - on the other hand - to **Z**, another trivial machine computing the subsequent internal state z'. From this it should be clear that the non-trivial machine too is synthetically deterministic.

I will have such a machine running for you in a moment, but would like first to get some terminology out of our way. **F** and **Z** are usually referred to as the Driving Function and the State Function respectively. Algebraically this is expressed by

$$y = \mathbf{F}\,(x,z), \qquad \text{Driving Function}$$
$$z' = \mathbf{Z}\,(x,z), \qquad \text{State Function}.$$

Perhaps you noted that the state function **Z** expresses a quantity (z') through itself at an earlier stage (z). This is the essence of recursive computations. I shall talk about these in point number (iii).

Let us construct now a minimal NTM, as closely as possible related to our TM of before. A minimal extension would be to add simply one internal state to that machine so that we have now instead of only one, two internal states. Let them be called I and II, and have the driving and state functions as follows:

When in I			When in II		
x	y	z'	x	y	z'
A	0	I	A	1	I
U	1	I	U	0	II
S	1	II	S	0	I
T	0	II	T	1	II

Now, let us explore the behaviour of this machine. I suggest testing first with the first input symbol A. We present the machine with several A's (A, A, A, ...), and to our satisfaction we get consistently zeros (0, 0, 0, ...). We turn now to a sequence of U's (U, U, U, ...), to which the machine responds with a sequence of ones (1, 1, 1, ...). Confidently we try the input S and obtain 1; but when checking out S again, for one who does not know the inner workings of the machine, something unpleasant is happening: instead of a 1, the machine responds with a 0. We could have predicted that, because the state function switches the machine when in I, given S, into its internal state II, and here the response to stimulus "S" is "0". However, being in II, given S, the machine returns to internal state I, and a new test of S will yield 1, etc., etc., ...

Checking out the patriotic sequence USA, depending upon whether one starts when the machine is in its internal state I or in II, it will respond with either 111, or else with 000; apparently indicating different political persuasions. Perhaps these examples suffice to justify the qualifier "non-trivial" for these machines.

More important, however, is to see the distinction between the one who knows the driving and state functions of the machine (may be he did the synthesizing), and the other one who has no access to this knowledge and is restricted to observing sequences of input/output pairs as his only base for hypothesizing about the inner workings of this machine.

At first glance, the distinction between the knower and the experimenter may appear to be not too severe. Clearly, the experimenter has the boring task of going through all these sequences to establish the rules that produce them; nevertheless, ultimately he should be able to crack the code of these machines, and their workings will become as transparent for him as for the knower: cumbersome, but possible.

Alas, this is not so.

Let me first turn to "cumbersome". The problem here is to identify among all possible machines with the given number of input and output states the one under investignation. By "identifying" is, of course, meant to infer from the observed sequences of input/output pairs the machine's driving and state functions.

Table **T**

1he Number of Effective Internal States Z, the Number of Possible Driving Functions $\mathfrak{N}_D$, and the Number of Effective State Functions $\mathfrak{N}_S$ for Machines with One Two-Valued Output and with from One to Four Two-Valued Inputs

n	Z	$\mathfrak{N}_D$	$\mathfrak{N}_s$
1	4	256	65536
2	16	2.10^{18}	6.10^{76}
3	256	10^{609}	$300.10^{4.10^3}$
4	65536	$300.10^{4.10^3}$	$1600.10^{7.10^6}$

In table **T** I have listed the numbers of the possible non-trivial machines with exactly two output states, say 0, 1, as is the case with ours, and with 2, 4, 8, 16, input states (n = 1, 2, 3, 4). Our machine has four input states A, U, S. T, (n = 2), hence our experimenter must search amongst

$$6.10^{76}$$

different machines to find the right one. Cumbersome? No! Transcomputational!

Now to "possible". There exists a large class of machines whose driving and state functions are such that it is *in principle* impossible to infer these functions from the results of a finite number of tests: the general machine identification problem is unsolvable! This also means that there are non-trivial machines that are unknowable.

I shall summarize now the essential features of non-trivial machines, and then conclude with a few comments. In parallel to what I have said earlier about trivial machines, one can say that all NTM's are:

   (i)   synthetically deterministic;
  (ii)   history dependent;
 (iii)   analytically indeterminable;
(iiii)   analytically unpredictable.

With the principle expressed in (iii) the non-trivial machines join their famous sisters who sing of other limitations:

Gödel: Incompleteness Theorem;
Heisenberg: Uncertainty Principle;
Gill: Indeterminacy Principle.

If one also takes the other unpleasantries of these machines into account, namely, the dependence on their past and their unpredictability, our efforts to remove or suppress all uncertainties in our environment are quite understandable. When we buy a machine we want it to function exactly as intended. When we turn the starter key in our car, it should start; when we dial a telephone number, we want the right connection, etc., etc., we want trivial machines. Hence, we like those guarantees that, in essence, are saying: "... at least for one year this machine will remain a trivial machine." If, in spite of this, it shows non-trivial tendencies (the car won't start, etc.) we call upon a specialist in trivialization who remedies the situation.

This is all very well. However, when we begin to trivialize one another, we shall soon not only be going blind, we shall also become blind to our blindness. Mutual trivialization reduces the number of choices, hence goes counter to the ethical imperative I voiced in the beginning. The task at hand is:

        de-trivialization.

3. Recursive Computations

Is the world a trivial or a non-trivial machine? Perhaps Einstein had an answer for this question when he said: "Raffiniert ist der Herrgott, aber boshaft ist er nicht" (Subtle is the Lord, but malicious He is not (15)). And Heisenberg: what would his answer have been after he saw that the interference of an observation leaves the observed in a state of uncertainty. Or should we switch his principle around and say more accurately that the interference of an observation leaves the *observer* in a state of uncertainty?

May be the original question contains an implicit flaw by stipulating a dichotomy between a world observed, and one who makes the observations. Perhaps each of us has first to answer for himself the question: "Am I a part of the Universe, or are we both apart." In other words, should we contemplate an epistemology in which I, the observer, am included in the domain of my observations, or shall we prohibit this re-entry (for ultimately we may see ourselves!).

Since the orthodox position here is to stipulate the separation of the observer from the world observed, a world usually perceived as a trivial machine whose function we are to uncover. Since this perspective is almost all-pervasive, I need not address myself to it.

Instead, I shall expand on the concepts of autology and closure of before, by making full use of the notions of "machines" whose behaviour under closure we are now to explore.

Consider an arbitrary large network of interacting NTMs that are fully connected to each other. By this is meant that each machine's output is an input for some others (or for itself); and each machine's input is an output from some others (or from itself) (see Figure 6a). Since there is no lead to the world outside of this network, this system is closed, it is its own world. Ross Ashby, who was one of the first to study the activity of such nets, referred to them as "systems without input" (16).

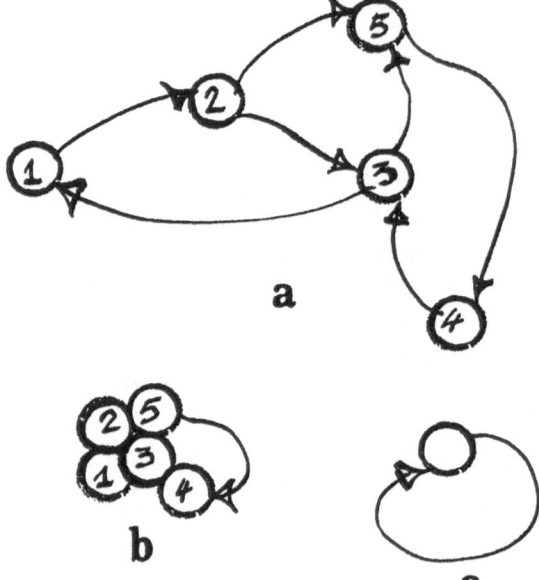

a

b

c

Figure 6: Network of Interacting Machines.

If we were to grab one of the connections between any two machines to observe the signal flow between them, it is irrelevant to how many more they are connected (6b): the whole net acts as a single NTM whose output is its input (6c).

Let us consolidate the operation of the entire net between the chosen points of in- and output into one operator

$$Op,$$

and let the result of this operation become the beginning of its next operation. In other words, let this become a recursive operation.

At this point I have struggled with myself whether I shall let you go through the paces of an elementary formal approach to recursive function theory, or whether I should make a short cut by just summarizing some results. Since I could not make up my mind, I decided to do both, for you can always skip various steps in the formal arguments, and turn to the summary. Nevertheless, I recommend you come along with me over the four points of this Primer on Recursions, because you will enjoy the consequences of the argument much more after having watched their development.

A Primer on Recursions

Elements of a Formalism.

1   Consider the (independent) variable x_0 (call it the "primary argument", being subscripted with "$_0$" to indicate that this is the variable taken ab ovo).

14

1.1 As the case may be, this variable may assume numerical values, or it may represent arrangements (e.g., arrays of numbers, vectors, geometrical configurations, etc.); functions (e.g., polynomials, algebraic functions, etc.); behaviours described by mathematical functions (e.g., equations of motion, etc.); behaviours described by propositions (e.g., the McCulloch-Pitts (TRE's temporal propositional expressions) etc.).

2   Consider an operation (transformation, algorithm, functional, etc.):

$$\text{"Op"}$$

acting upon the variable x_0; indicate the action on this operand x_0 by

$$Op(x_0)$$

Call x_1 the values generated by the first application of Op on x_0

$$x_1 = Op(x_0) \qquad (1)$$

or graphically

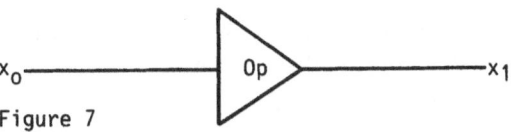

Figure 7

2.1 Apply Op to x_1, and call x_2 the values so generated:

$$x_2 = Op(x_1), \qquad (2)$$

that is, x_2 represents the values generated by having Op applied twice to x_0. (With Equ. (1) and (2)):

$$x_2 = Op(x_1) = Op(Op(x_0)). \qquad (3)$$

2.2 Call $Op^{(n)}$ the n-th application of Op to a variable, then:

$$x_n = Op^{(n)}(x_0), \qquad (4)$$

or graphically

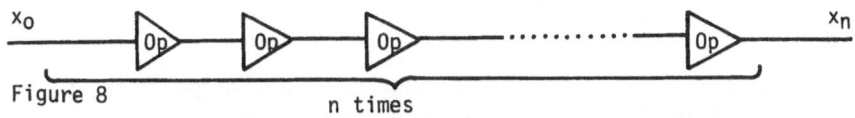

Figure 8   n times

3   Consider the case in which Op is applied indefinitely ($n \longrightarrow \infty$) to a variable, say, x_0:

$$x_\infty = Op^{(\infty)}(x_0) \qquad \text{or} \qquad (5)$$

$$x_\infty = Op(Op(Op(Op(Op(Op(Op(Op(Op(Op(Op(Op(Op(... \qquad (6)$$

3.1 Contemplate expression (6) and observe:

3.11 That the independent variable x_0, the "primary argument" has disappeared;

3.12 That, since x_∞ expresses an indefinite recursion of the operators Op onto operators Op, any indefinite recursion within that expression can be replaced by x_∞:

15

$$x_\infty = Op(\ Op(\ Op(\ \ldots\ldots\ldots$$

(diagram showing recursive substitution of x_∞)

3.2 Hence:

$$x_\infty = Op(x_\infty) \tag{7.1}$$
$$x_\infty = Op(Op(x_\infty)) \tag{7.2}$$
$$x_\infty = Op(Op(Op(x_\infty))) \tag{7.3}$$
$$\text{etc.}$$

3.3 If there are values $x_{\infty i}$ $(i = 1,2,3,4, \ldots m)$ that satisfy equations (7), call these values

"Eigen-Values"
("Self-Values")

$$E_i = x_{\infty i}$$

(or Eigen-Functions; Eigen-Operators; Eigen-Algorithms; Eigen-Behaviours (= "Objects") etc., depending upon the type of the primary argument).

4 Contemplate expressions (7) and observe:

4.1 Eigenvalues are discrete (even if the primary argument is continuous). This is so, because any infinitesimal displacement $\pm\epsilon$ from a stable Eigenvalue E_i (i.e. $E_i \pm\epsilon$) will disappear, as do all values of x_0, except those for which it happens that $x_0 = E_i$.

4.2 Closure:

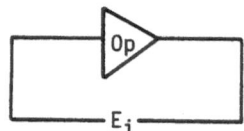

for only under this condition are operand and operatum equivalent. That is:

4.21
$$\lim_{n \to \infty} Op^{(n)} = Op \tag{8}$$

4.3 Since an operator implies its eigenvalues E_i, and vice versa, operators and eigenvalues are complementary ($Op \longleftrightarrow E_i$; they may stand for each other).

4.31 Since eigenvalues produce themselves (through their complementary operators), eigenvalues are self-reflexive.

Examples

1.) Consider the array

$$1,2,3,4,5,6,7,8,9,0\ .$$

Apply to it Ashby's "Evolutionary Operator" EV:
EV = "Choose two numbers at random; from the (two-digit) product (e.g. 2x3 = 06); replace the two chosen numbers by the digits of the product".

$x_0 =$ 1,$\underline{2}$,$\underline{3}$,4,5,6,7,8,9,0

$x_1 =$ 1,0,$\underline{6}$,4,5,$\underline{6}$,7,8,9,0

$x_2 =$ 1,0,$\underline{3}$,4,5,6,$\underline{7}$,8,9,0

$x_3 =$ 1,0,2,4,$\underline{5}$,6,1,$\underline{8}$,9,0

$x_4 =$ 1,0,2,$\underline{4}$,4,6,$\underline{1}$,0,9,0

$x_5 =$ 1,$\underline{0}$,2,0,4,6,4,0,9,0

(observe the vanishing of odds)

$x_6 =$ 1,0,2,0,4,$\underline{6}$,4,0,0,$\underline{0}$

$x_7 =$ 1,0,2,$\underline{0}$,4,0,4,0,$\underline{0}$,0

(observe the emergence of naughts)

$x_8 =$ 1,0,2,$\underline{0}$,4,0,4,0,$\underline{0}$,0

$\vdots$

$x_{15} =$ 0,0,2,0,4,0,0,0,0,0

$\vdots$

$x_\infty =$ 0,0,0,0,0,0,0,0,0,0 $= E_1$

2.) Consider the array

1,2,3,4,5,6,7,8,9,0 .

Apply to it Ashby's "Co-Evolutionary Operator" CE:
CE = "Choose two numbers, α,β, at random; change β to the last digit of

$$\alpha^4 + \beta^4;$$

leave α unchanged.

From the following Table, which lists these last digits for each of the pairs, α, β, one may convince oneself that the Eigen-Arrays either contain 2's and 7's in equal numbers, or else 2's only. (Note that in case the 2's disappear completely, they will be regenerated through the 7's. The converse is not true).

Table

	1	2	3	4	5	6	7	8	9	0
1	2	7	2	7	6	7	2	7	2	1
2		2	7	2	1	2	7	2	7	6
3			2	7	6	7	2	7	2	1
4				7	1	2	7	2	7	6
5					6	1	6	1	6	5
6						2	7	2	7	6
7							2	7	2	1
8								2	7	6
9									2	1

3.) Consider the operator "Taking the Square Root" SQR, and apply it recursively to an arbitrary initial value x_0.

The attached Table gives the print-out of the sequence x_1, x_2, x_3, ... etc., for the initial value:

$$x_0 = 137.$$

Observe the convergence to the Eigenvalue

$$x_\infty = 1 .$$

Observe also the complementarity

X'=SQR(X)
INITIAL X = 137

11.70469991	1.00965564	1.00003753	1.00000014
3.42121322	1.00481622	1.00001876	1.00000007
1.84965218	1.00240521	1.00000938	1.00000003
1.36001918	1.00120188	1.00000469	1.00000001
1.1661986	1.00060076	1.00000234	1
1.07990675	1.00030033	1.00000117	1
1.03918561	1.00015015	1.00000058	1
1.01940453	1.00007507	1.00000029	1

4.) Consider the two operators "Cosine" and "Sine" operating onto each other recursively:

$$x' = \cos(y)$$
$$y'' = \sin(x').$$

The attached Table gives the print-out of the sequence

$$x_1, y_1, x_2, y_2, x_3, y_3, \ldots \text{ (in radians)}$$

for the initial value

$$y_0 = 3 \text{ rad.}$$

Observe the oscillatory approach to the Eigenvalues of the two operators "seeing themselves through the eyes of the other":

$$\cos(\sin(0,768169..)) = 0,768169..$$
$$\sin(\cos(0,694819..)) = 0,694819..$$

Note the difference of the Eigenvalues of these operators, when each operator is taken separately:

$$\cos(0,739085..) = 0,739085..$$
$$\sin(0,000000..) = 0,000000..$$

Observe also the rapid convergence to mutual Eigenvalues. After only 36 steps the stable values are approached within one in a million.

INITIAL Y=3

-0.9899924293	0.6916683255	0.7681274735	0.6948203121
-0.8360218258	0.7701829943	0.6947897149	0.7681687568
0.6704198624	0.6962666018	0.7681883513	0.6948194033
0.6213150305	0.7672419786	0.6948334981	0.7681693438
0.8131136789	0.6941525818	0.7681603226	0.6948198267
0.7264305416	0.7685961014	0.6948133393	0.7681690722
0.7475500224	0.6951266802	0.7681732221	0.6948196347
0.6798440992	0.7679725702	0.6948226158	0.7681692011
0.777670743	0.6946783	0.7681672922	0.6948197269
0.701621614	0.7682596786	0.6948183524	0.7681691408
0.7637965103	0.6948847942	0.7681700157	0.6948196821

I hope that with this brief description of some points in recursive function theory, and with the few examples of its application, you could, at least, get a flavour of this method, and could see in the recursive operation a principle of self-
organization that allows certain structures to emerge - to crystallize - from early, arbitrary stages. However, many other interesting results I have not mentioned,
results involving multiple eigenvalues, compositions of such states, and many more.
Moreover, examples in which the eigenstates are not numerical quantities, but are
themselves operators (Eigen-Operators), or of other domains would have been illumi-
nating. This would, however, require a much more elaborate formal apparatus, and
for the study of such cases I have to refer to the literature (11) (17) (18).

Nevertheless, I cannot leave this account without a short note on the results of
Ashby's studies of the dynamics of large systems without input. In a computer simu-
lation of an arrangement as in Fig. 6a, Ashby connected in one series of experiments 100, in another 1,000 non-trivial machines (essentially computing on their
inputs a variety of logical functions), and after setting them at a initial value
let them loose. After some transients in the beginning (see also our examples) the
systems settled into various eigenbehaviours, i.e. "limit cycles", of different
length, in many cases representing large domains of initial values. Polystability
was his term for this phenomenon (16). His studies have recently been revived with
much faster and larger computers by a French group leading to many new and fascinating results (19).

I shall now conclude my story on recursive computation with a few words on terminology.
As I mentioned before, it was David Hilbert who, around the turn of the century,
introduced the terms Eigen-value and Eigen-function, terms I find particularly well
chosen for representing the logic that is involved here. Somewhat later, another
attractive feature of these values, namely the invariance under their corresponding operations, brought them the name "fix points". And recently, some computer
buffs discovered these fascinating values for themselves, and since they could not
believe their eyes when they saw what they saw, they called these values "strange
attractors", a term, I am sorry to say, I find repulsive.

4. Socio-Managerial

Malik and Probst in their article on evolutionary management looked, of course,
upon the role of negotiation from the perspective of the firm as an evolving,
self-organizing system. I would like to supplement their observations with points
that emerge from the strategies I just reported.

I propose to look for the moment at negotiation as an attempt by members of a group
to "solve a common problem". The quotation marks here I intended to be acting as

flags, as caution signs, inviting us to re-examine the over-used and over-abused terms so quoted. What is meant by "solve", by "common", by "problem"? Most likely, there is no common problem! each one of the members may have his own; worse, may be he does not *have* a problem, perhaps *he* is the problem; etc.

With this warning in the back of our minds, I propose again to look at negotiations as a problem solving task, where one of the solutions may indeed be the identification of a "common problem".

Small Group Dynamics

I will describe now one of the early experiments in small group dynamics, an experiment which is to my taste much too little known for the many interesting conclusions one can draw from its results. This experiment was designed in the early 'fifties by Alex Bavelas (20), then at M.I.T., who became interested in the evolution of strategies and feelings of people with different expertise, who participate in various tasks whose ends and means are given in terms that span from transparency to opacity. There seems to be a similarity here with the situations in which the Principle of Potential Command may have its application. This, however, is not the case, for here intended actions are controlled in a way, as we shall see in a moment (and for the record monitored).

The task given to the members of a group of five is to find the only common symbol in a deck of cards, of which each member has only one card to look at, but can communicate with others, exclusively through prescribed channels, to get the needed information for the other symbols. Let me first describe the deck of cards, and then the spatial arrangement of this experiment.

Cards: Consider six different symbols, say, a square, a cross, a triangle, etc., which I conveniently will label **1,2,3,4,5,6,**. Design six different cards, each with one symbol missing, but showing all other five:

1 2 3 4 5 6
The last line of this schema indicates the missing symbol.

It is clear that by removing from this set one card that lacks, say, symbol **3**, the remaining cards of the set will have one, and only one, symbol in common, namely **3**.

It is also clear that in this way six sets, or decks, of cards can be generated, each deck distinct from the others by its common symbol.

In a preliminary "get at ease" session each prospective participant is given such a set with the question to identify the common symbol. That takes between 20 and 40 seconds to answer. At that time he is also told that in the actual experimental situation he will see only one card, and has to infer the common symbol through interactions with other members of the group.

Space: Consider two concentric pentagonal cylinders, where the space in between them is subdivided into five identical compartments, each of which can seat comfortably one of the five participants. In front of the wall facing the centre is a wide and shallow desk. In the wall above the desk are slots, some open, some closed, through which messages can be sent to, or received from, other participants via tubes that are concealed behind the walls. Communication through these tubes is the only way participants are able to interact, sound proof walls, etc., restricting other means.

Two crucial points in the design of the experiment are (i) the possibility of specifying beforehand (unbeknown to the participants) the connectivity between compartments, for instance, those of Figures 9; and (ii), the possibility of keeping track of the communication process through numbered and colour coded message pads and pencils.

Experiment: A single session begins with the five participants seated in their compartments and facing one card in the set of five. They may use their note pads for any message, a question, an answer, whatever, being sent to others. As soon as each participant thinks he knows the common symbol, he presses the appropriate key, one of six on his desk. The session ends when all participants have pressed the same key.

Results: Although the experiments yielded a large crop of results, I shall talk only about two kinds of variation within this overall design. One regarding connectivity, the other different symbols. Variation in connectivity, say, from "circle" (Figure 9a) to "star" (Figure 9c), produces changes in performance that are already quite impressive. When varying symbols, the changes were dramatic. In one set of experiments identifiable symbols were used in groups connected differently. In the other set "noise" was introduced into the communication channels - or should I say into the channels of cognition - by using "symbols" that are not only difficult to distinguish, they don't even have names: here sets of differently mottled marbles were used in lieu of symbols.

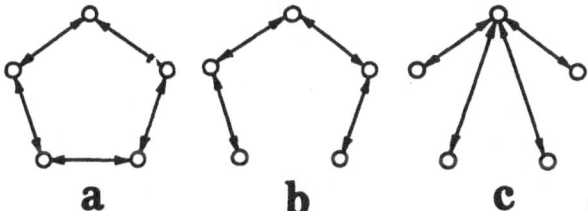

a   **b**   **c**   Figure 9: Various Connected Groups

Let me report first about the "noiseless" case, i.e. the one in which the symbols were identified and named. When those who played the "circle" were asked how they felt during the session, how they perceived their performance, etc., they consistently replied that they were feeling fine, that they performed fast and efficient, that they may have done better, etc. When asked whether they could identify a "leader" in the group, the averaging of replies distributed this "leadership" evenly over all five positions.

For the "star" performers the story is almost the opposite of the "circlers". Although their performance is about twice as fast than that of the circle groups, they had the feeling of defeat. They sensed themselves as slow and inefficient. They blame some "idiot" in the team for that. Of the participants 94 % identify the apex of the connectivity to be the leader.

Because of the difference of perceiving the absence or the presence of a leadership, Bavelas and his co-workers nicknamed these two connection schemes "democratic" and "authoritarian".

What happens now when "noise" is introduced? Surprisingly (or perhaps not surprising at all), the democratic group works just as well, however somewhat slower than before. They still feel fine and think they are doing well. The dramatic change is with the authoritarians: depending on the "strangeness" of the symbols, the groups disintegrate sooner or later. Participants walk out in anger, the "idiots" multiply, and blame is passed from one to others. Indeed, when later the communication records are studied, the star performers soon stop talking about symbols, they start calling each other names. There is a fascinating switch from an attention to communicabilia to that of the communicators.

The difference to "democrats" is fundamental. Expanding language is what keeps the people in this configuration going. As the records show, names for the funny looking things are soon invented, some referential, "lion", "cow", etc., or new ones "splops", "bimbim", etc., names that are either kicked around, modified, or kept; and when adopted by the group, the find-the-common-symbol problem is back to finding common *symbols*, by ignoring fuzzy objects.

I went through these exercises at a somewhat greater length, for I felt that these experiments are superbly suited for connecting the four notions, management, self-organization, evolution, and language, very much in the sense of Malik and Probst admonished us in their paper.

There is no doubt that one of the managerial tasks is to generate a climate that fosters communication. One of the outcomes of the Bavelas experiments suggest that interaction structures can be facilitators or inhibitors for communication. It apears that circulatory, recursive interaction patterns are highly stable against perturbations. The important point, however, is that this stability is not through counteracting the perturbing forces, but by utilizing them as a wellspring of creation. And finally, these experiments show again the significance of language in the managerial process (21).

My friend, the composer Herbert Brun, once taught me "...a language learned is a language lost" (22). But here, in one of Bavelas' situations, are instances of language in the make.

Lexically speaking, language is a closed system: ask for the meaning of a word, and you get words. I want to know the meaning of "subsequent". The dictionary (1) tells me "following" (see Figure 10). I want to know what this means, etc. ... Figure 10 tells where this all leads to; one may say to nowhere. Can one get out of this trap?

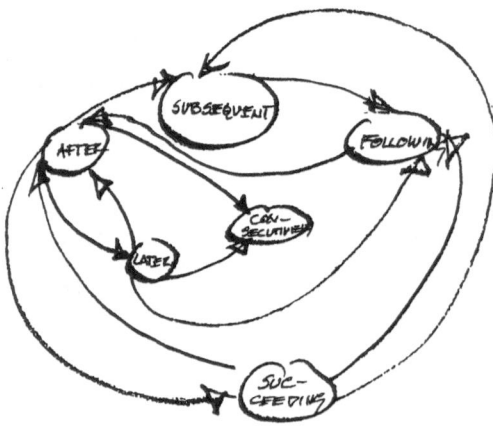

Figure 10: Relational Network of Synonymos
Terms.

I suggest one path that was seen by the British philosopher John Langshaw Austin. He observed in our language a peculiar family of utterances that say what they do; or perhaps I should say, they do what they say. Now, how is that?

Imagine yourself in a crowded bus; inadvertently you step on somebody's toes. Politely you say:

"I apologize".

The magic of this utterance is that it *is* the apology. For obvious reasons, Austin called these utterances "performative" (23). Once one is aware of these utterances in our language, one sees them appearing more and more: "I promise", "I declare" .., etc. Contemplate for a moment the extraordinary things that are going on when in a marriage ceremony the priest says:

"I declare you husband and wife".

When this formula is uttered, they *are* husband and wife.

The notion of the performative utterance I brought up at the end of my story, for - in proper recursive fashion - it ties me back to the beginning. You may remember the sentence that says of itself how many letters it has. When indeed it says so correctly, we called it an Eigen-value. May be, one should call it an Eigen-utterance, to let the connection with performative utterances become visible. Here, I suggest, is the window through which we can step outside of language. Hence, let me conclude with a reference to your kindness for having invited me, and to your patience in listening to me, in form of a performative utterance:

"Thank you very much!"

References

(1) The American Heritage Dictionary of the English Language, Houghton Mifflin, Boston (1966).

(2) Malik, F. and G.J.B. Probst: "Evolutionary Management" Cybernetics and Systems: Int. J., 13, 153-174 (1982).

(3) von Foerster, H.: On Constructing a Reality" in Observing Systems, A Collection of Papers by Heinz von Foerster, Intersystems Publications, Seaside (1982).

(4) Gunther, G.: "Cybernetic Ontology and Transjunctional Operations" in Beiträge zur Grundlegung einer operationsfähigen Dialektik, I, Gotthard Gunthers gesammelte Werke, Felix Meiner Verlag, Hamburg (1976).

(5) Personal communication from Professor Lars Loefgren, Dept. for Automata Theory and General Systems, Building E, University of Lund, Box 725, S-220 07 LUND, Sweden.

(6) Varela, F.G.J.: "A Calculus of Self-Reference" Int. J. General Systems, 2, 5-24 (1975).

(7) Psk. G.: "The Meaning of Cybernetics in the Behavioural Sciences" in Progress in Cybernetics, J. Rose (ed.), Gordon and Breach, New York, 1, 15-44 (1969).

(8) von Foerster, H.: "Foreword" in Rigor and Imagination, Essays from the Legacy of Gregory Bateson, C. Wilder-Mott and John H. Weakland (eds.) Praeger, New York (1981).

(9) Wittgenstein, L.: Philosophical Investigations, G.E.M. Anscombe (tr), The Macmillan Company, New York (1953).

(10) Hofstadter, D.R.: "Metamagical Themas" Sc.Am.Jan., 12-332 (1981), and Jan., 16-28 (1982).

(11) Davis, M.: Computability and Unsolvability, McGraw-Hill, New York (1958).

(12) McCulloch, W.S.: "A Heterarchy of Values Determined by the Topology of Nervous Nets" in Embodiments of Mind, MIT Press, Cambridge (1965).

(13) Turing, A.M.: "On Computable Numbers, with an Application to the Entscheidungsproblem" Proc. London Math. Soc., ser. 2, 42, 230-265 (1936-1937).

(14) Gill, A.: Introduction to the Theory of Finite State Machines, McGraw-Hill, New York (1962).

(15) Pais, A.: "'Subtle is the Lord ...' The science and the Life of Albert Einstein", Oxford University Press, New York (1982).

(16) Walker, C.C./Ashby, W.R.: "On Temporal Characteristics of Behavior in Certain Complex Systems" Kybernetik, 3 (2), 100-108 (1966).

(17) von Foerster, H.: "Objects: Tokens for (Eigen-)Behaviors" in Observing Systems (see Ref. 3.).

(18) Hofstadter, D.R.: "Metamagical Themas" Sc.Am. Nov., 22-43 (1981).

(19) Fogelman-Soulie, F., Goles-Chacc, F. and G. Wissbuch: "Specific Roles of the Different Boolean Mappings in Random Networks" Bull. Math. Biol., 44 (5), 715-730 (1982).

(20) Bavelas, A.: "Communication Patterns in Problem-Solving Groups" in Cybernetics, Heinz von Foerster (ed.), Josiah Macy Jr. Foundation, New York (1952).

(21) Request (and indeed request!) literature from: Hermenet, 1750 Union Street, San Francisco, California 94123; Attention Dr. F.C. Flores.

(22) Brun, H.: "Futility 1964" (side 5, band 3) in Compositions by Herbert Brun, Non Sequitur Records, Box 872, Champaign, II 61820 (1983).

(23) Austin, J.L.: "Performative Utterances" in J.L. Austin: Philosophical Papers. J.O. Urmson and G.J. Warnock (eds.), At the Clarendon Press, Oxford (1961).

Two Principles for Self-Organization

F. Varela

Departamento de Biologica, Facultad ciencias basicas y farmaceuticas
Universidad de Chile, Casilla 653, Santiago, Chile

1. Principles

In this paper I would like to propose two guiding principles for the study and
understanding of self-organization in natural systems, which are derived from a
combination of both empirical observation and cybernetic considerations. The basic
thrust of these ideas is to see self-organization as a behavior of a specific class
or type of systems, whose organization can be clearly spelled out. This amounts to
explore the underlying mechanisms for self-organization itself. The two principles
are as follows:

Principle 1:

Every operationally closed system has eigenbehaviors.

Principle 2:

Every operationally closed system changes by natural drift.

Obviously, as stated, these two principles are opaque. The remainder of this paper
will attempt to unfold their meaning.

By operational closure I mean a class of "organizations". In fact, every system,
once distinguished through a certain criterion, has two complementary aspects: its
organization, which are the necessary relations which define the system, and its
structure, which are the actual relations between the components which integrate
the system as such (1). Thus, ex definitio, the organization is invariant while a
system maintains its identity without disintegration; structures can vary provided
they satisfy the organizational constraints.

In this light, the key step in defining a system is spelling out its organization.
My view on this is that only too often in cybernetics we have been misled by the
positivistic assumption that a system *is* something or other, and hence that its
organization is unique.

Instead, since we distinguish a system according to a criterion of distinction
given by an observer, I argue that there are many possible ways of specifying an
organization for a given distinction, depending on what the observer *considers*
satisfactory or convenient. This point is far from idle. One of the most important
alternatives in characterizing a system's organization is the difference between
an input-type description, and a closure-type description.

By an *input-type description* I mean that the definition of the system's organization is given by the *specific* ways in which it interacts with its environment,

through a well-defined set of inputs followed by a transfer function. This is no surprise since it has been the standard mode of description in system's theory and cybernetics. It entails a specific concept   of what a mechanism is, and a corresponding mode of inference.

However, the study of biological system forces us to consider a complementary mode of description which is based on the fact that some systems exhibit, intuitively speaking, an internal determination or self-assertion. For such autonomous systems, the main guideline for their characterization is not a set of inputs, but the nature of their *internal coherences* which arise out of their interconnectedness. Hence the term *operational closure* (2).

The main consequence of changing from an input- to a closure-type stance for defining the organization of a system is that we concentrate on the inner coherences, and thus what used to be specific environmental inputs are bracketed as unspecific perturbations or simply noise. An input becomes a perturbation when it is no longer necessary to specify the system's organization, i.e. it has become noise.

Examples of organizational closures abound: nervous systems, immune systems, ecologies, conversations, ... in all these cases the degree of self-involvedness of the system is such that by and large the most important part of what needs to be studied are the qualities of this self-involvement. To be sure, there are some limiting factors or *constraints* which the system must satisfy. But within the ample range of satisfying these constraints, all interactions are treated as noisy perturbations. This amounts to dealing with a different kind of mechanism, and a correspondingly different mode of inference.

Please note that I am putting the emphasis on what is necessary to come up with a satisfactory explanation to *us*, not in the sense of any ontology of systems-out-there. My plea is not for exclusivity, but for clean epistemological accounting. That is: to be fully aware when and why we are using one type of description, and when we change from one to the other. And when it comes to biology, it becomes clearer and clearer that closure has to be taken seriously if we are to make any headway in the study of some of their most interesting qualities: cognition, innovation, learning, evolution.

Now, one of the most striking experiences of the study of systemic closure is the ease with which they exhibit a coherent behavior emerging from their mutually interdependent components. In fact, the apparently trivial random networks first studied by Ashby and von Foerster in the 1950's were the first timid announcement of this surprise. It is as if once the closure of the system is achieved, it automatically takes care of the generation of its internal regularities. I call such internal coherence *eigen-behaviors* for obvious reasons (3).

Recently, the interest in the study of the deeper reasons for these empirical observations has been rekindled. Thus, for example, we have now the beginnings of a classification of the eigenbehaviors of boolean networks (4). But a robust theory, comparable to the theory of Turing automata (i.e. of input-type descriptions), is still far away from us, and it depends partly on a re-evaluation of the role of self-reference altogether. I do not want to go into that here, but simply to take it at face value.

The unfolding of what was contained in *Principle 1* is thus complete. Loosely stated says: if it is meaningful to characterize a system as network-like, it will necessarily produce a landscape of internal coherences.

This, I believe, is the crux of self-organization, and therefore Principle 1 serves to explicate or even define what a self-organizing system is. Properly speaking there are no self-organizing systems, only self-organizing *behaviors*, which are pointers to the need to achieve an explicit characterization of the organizational closure of a given system underlying such a behavior.

Now as to *Principle 2*. The question that Principle 1 evades completely is *how* does such a system couple or interact with its medium so that it maintains its identity. Furthermore, how is it that natural systems seem to relate to their environment in ways which are adaptive. This question, of course, has a classical answer from an input-type description: adaptation is optimization to given inputs. Since the system is defined through its inputs, there is no problem in giving sense to such an optimal match. But when we have bracketed the inputs, so that they become per- turbations, and have reversed our mode of inference in the opposite direction, from coherence to perturbation from the medium, the question does not have a cano- nical answer.

Again here it is essential to recall our context. Remember that a system maintains its identity to the extent that it maintains its organization. Its structure never- theless can vary substantially, and therefore the kinds of eigenbehavior available to the system at every moment of time will vary (by Principle 1). From an input type stance, a system is adaptive because it is optimally fitted to a given world. From a closure type stance, a system is adaptive simply because *its organization is maintained invariant* through changes of structure which do not violate con- straints. In the first case we have the relation system/environment as an instruc- tive or prescriptive rule. In the second, this relation becomes a satisfying or proscriptive rule. As long as the identity is maintained through the conservation of the organization, there is adaptation, but it does not say in which particular way this can come about. The ways are dictated by the system's closure and the corresponding eigenbehaviors.

Thus, in the history of coupling of a system with its medium there will be an un- interrupted synthesis of eigenbehaviors which are specified by the system itself, under the broad constraints of the environment. Hence the possible landscapes of such eigenbehaviors are multiple. I have chosen the name *natural drift* to designate such multiple paths of identity invariance (5).

Thus explicated, the second principle tells us how an operationally closed system, rich in eigenbehavior, can actually endow a world of perturbations with a *meaning* as its own world. Obviously, here I have not touched upon an essential topic, name- ly, what are the *rules* for structural changes which lead to maintained viability. This is another line of research for which, again, there is tremendous current interest, but that I will not consider here any further.

Enough of the overall settings. I would like now to consider briefly how does this change in optics lead us to a different and pragmatically distinct way of under- standing brain functioning and evolutionary phenomena.

2. Brain and Evolution

Brain processes and evolutionary phenomena have been the examples of self-organi- zation par excellence. That is, a situation typical of individual ontogeny, and a situation proper to a supra-individual unit, a population. Let me sketch how a closure-type stance to both these situation changes quite radically our approach to their study. In this sense, these two examples are my evidence empirically to substantiate the two Principles as proposed above.

An input-type stance for the *brain* amounts to treating the nervous system as ope- rating fundamentally on the features of qualities of the environment which should be taken up as the raw material to be processed inside. In brief: the nervous system works with a *representation* of the information content from the ambient.

The closure type stance amounts to treating the nervous system as being defined fundamentally through the internal coherences which are attained by its relative interconnectivity. More precisely stated, by the eigenbehaviors which are produced through the mutual (neuroanatomical) *mappings* between its various internal surfaces together with the local actions proper to each one of them. Sensory and motor actions are, from this point of view, only one link in an ongoing operational closure, albeit an important one for the observer of behavior (6).

The reader will agree with me that these two characterizations amount to seeing rather different systems, and to look for rather different things in the experimental designs. By and large the representational view is more predominant today, (and it has been over the last fifty years). Yet, there is the closure-type alternative view which seems, to me, both more interesting and more simple. As I cannot develop in full this point of view, I will have to make do with a few remarks to motivate it, lest the reader thinks it is totally crazy to say that the nervous system works without inputs and outputs. The representational view needs no motivation, since all text books are built around it.

Consider the realm of visual experience. It seems natural, at a first glance (so to say), to agree that the world appears as full of textures which by simple actions, like moving my head, are neatly taken in by my perception. This first-order experiental view seems to suggest naturally an input-type stance and looking for features of the environment which are picked up by various stages in my retina, then on up to the visual cortex.

However, this first impression of our experience is misleading. A closer look reveals that indeed what I see has more to do with the way I am put together as a mechanism than with an out-there. This more subtle view of experience is in fact easier to correlate with the structure of the nervous system. Thus, for example, whatever happens at the level of my retina as a sensory surface has only one effect: modulation of neural activity in the several neuronal aggregates to which it connects on a retinotopic basis. One of them is the lateral geniculate nucleus (LGN) of the thalamus, which the text calls "a relay station to the cortex". However, as one fiber from the retina comes in at a certain position in the LGN, at least five others come in at the same location. As a result, whatever the effect on the retinal activity is, it cannot but act as only a modulator of something which is ongoing inside the nervous system. In this case what is relevant is the state of relative activity between the various interconnections of the LGN with other brain layers, and through them, with the whole brain. A diagram may help to visualize the situation:

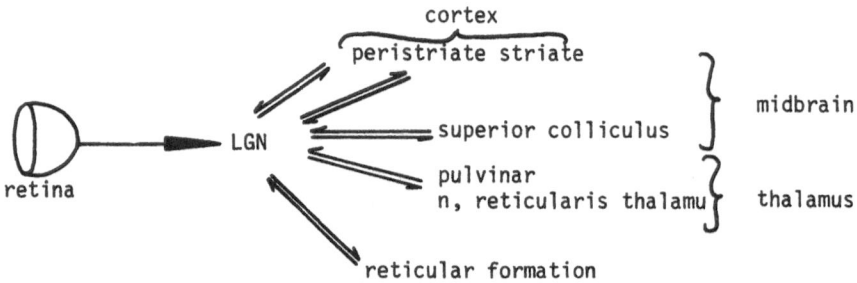

This kind of system is more naturally (or more easily) described as one whose internal coherence, attained through the reciprocal layer-to-layer interconnections, is *modulated* via a coupling surface, such as the retina. But the key to the system's operation is the attainment and diversity of its eigenbehavior, not the nature of the modulation. For example, we tend to think of color as an attribute of objects.

However, on closer examination color is virtually independent (except in very restricted situations) from the illumination that reaches the eye. Color is defined for us, experientially, through a mechanism to which we have no direct experiential access. Such mechanism essentially consists of an operation of relative comparison between levels of activity, and the invariants obtained through this sort of mechanism *do* correlate well with our experience of colors (7).

If we shift our stance towards the nervous system from an input-type to a closure-type, then the possible mechanisms of learning and memory also undergo a corresponding change. One of the universal properties of all known nervous tissues is that their structure at the local and cellular level is plastic. The conclusion immediately follows: given a system with operational closure, whose structure changes slightly under a history of perturbations, to an observer it will look (as if it had) learned, and that it keeps a record of what has happened. For the system however, no such record is necessary, for a given change can happen at many sites simultaneously and has no direct relationship with the perturbing agent. It is only for the observer that, if the perturbation is recurrent, this event will look like a recognition. Computers work by storage and retrieval; nervous systems do not, they learn by their natural drift.

This doesn't mean that we can say now with accurate precision exactly how the landscapes of eigenbehaviors change with the small scale changes at the cellular synaptic level. Work in this direction is just beginning. We need a better understanding of how to characterize neural eigenbehavior, and how new eigenbehavior can arise under perturbations. This is certainly a major area to work in (8).

Let me now turn to a corresponding characterization of the two viewpoints about evolution which amount to taking either an input or a closure-type stance.

The *input-type stance* amounts to treating the environment as the main guideline to understand the dynamics of the modifications by descent and its genetics, the result being a better *adapted* lineage of organisms. In other words, natural selection is an example of an optimization algorithm: the survival of the fittest.

The *closure-type stance* amounts to treating the internal coherences of an animal as the guiding trait to understand phyletic transformation. The result is that there is the generation of diversity but no sense of optimization of adaptation. In fact, adaptation is an *invariant*, just like identity is an invariant as long as the organism does not disintegrate (9). Natural selection is (in this case) here only a description of the major boundary conditions in which range such phyletic diversity can proceed.

The reader will, again, agree with me that these two stances amount to rather different views of evolution, and that they imply rather different things to look for and in designing experiments. Interestingly enough, the input-type stance or adaptationist paradigm has also been in dominance over the past fifty years (at least in the anglo-saxon science). Yet there is today, and there has been for quite some time, (in fact, since Darwin himself) the alternative closure-type stance. Again, the textbook view needs no motivation, for we are all used to it. Lest the reader thinks, however, that to abandon an adaptationist viewpoint is as suicidal as abandoning a representational view of the brain, let me say a few words of justification for this view.

Consider the bodyplan of any organism. The adaptationist tendency is to see this body as a collection of traits whose design can be explained though an optimal fit with the corresponding aspects of the environment. At best such traits are seen as having some cross correlations where a cost-benefit analysis can apply. However, this result is still a collection of traits. It is thus that we have heard the stories on how did the fish acquire its fins and the kangoroo its big feet: it is a superior design for what it is meant for.

However, this first impression is, again, misleading. For it leaves out the most striking fact about both development and organismic physiology, namely, that the unit does not work as a set of traits but as a coherent whole. The position of a limb in an embryo cannot be understood as some natural provision for what is to come next but as the result of the interdependence and mutual definition of what goes on in an *embryo* at each point. To change one point is to change radically its eigenbehavior, and not just one trait (10).

Thus an important school holds that the constraints in a bodyplan which arise from this operational closure of an organism "restrict possible paths and modes of change so strongly that the constraints themselves become the most interesting aspect of evolution" (11). For example the beautiful and intricate patterns of the shells of certain molluscs are most easily explained as diversity which arises from the diversity of materials to satisfy an invariant mode or architectural development. Correspondingly, these different forms do not correspond to some sought-for optimal adaptation (12).

This is not to say, again, that we have all the details of how such mechanism works. In fact, at this stage, evolutionary theory is in the middle of a profound revision. However, independent of the kinds of explicit mechanisms that are refined to express these ideas, it is evident that the change of our stance from one of optimum design and adaptation, to one of evolution as natural drift, is a very profound one.

3. Conclusion

Let me present the bare skeleton of the logic I have followed. It consists of five links:

(1) Self-organization is a behavior which is proper to autonomous units;

(2) autonomous units can be appropriately characterized if we change from an input-type to a closure-type stance;

(3) specifying the closure of a system leads to an understanding of the internal coherences (eigenbehaviors) such units have (Principle 1);

(4) if a system has enough structural plasticity the landscape of its eigenbehaviors will be diverse and complex, and the pathways of change from one to another will be constrained, but not uniquely specified: there is a natural drift (Principle 2);

(5) such self-determined internal coherences and their natural drift, when observed as behaviors under contingencies of interactions, will appear as the making of sense, novelty, and unpredictability, in brief, as the "laying down" of a world.

This reasoning constitutes, in my opinion, an *explication* of the mechanism of self-organization because it is capable of *generating* it. In the second part of this paper I discussed briefly how this leads to a change in our current understanding of at least two important phenomena: brain operation and evolution.

I wish to conclude with two commentaries.

First

It is no curious coincidence that for the last few decades, in two major areas of biology - neurobiology and evolution - an input-type description was chosen instead of a (perfectly accessible) closure-type stance. I believe this touches on a very deep prevalent contemporary sensibility of our contemporary science. The best way I know of formulating it is by saying that it favors a *representational epistemology*, that is, the predominant view of *knowledge as a world picture*. This assumption is familiar, especially in the context of anglo-saxon scientific philosophical tradition.

In contrast, I firmly believe that there is a major change, or trend of a change in our contemporary sensibilities and scientific epistemology in the sense that we are becoming more and more interested in an epistemology which is not concerned with the world-as-picture, but with the *laying down* of a world, where a unit and its world co-arise by mutual specification.

It could be said that the notion of self-organization serves as a clear symptom that differentiates between input-type machines (whether we call them Turing, state-transition, or simple functionalist theories), and biological autonomy and human understanding. This is so because self-organizing behavior depends on a history of coupling and it is based on a mechanism which is explicitly distributive and hermeneutic: it is interpretative, precisely, in the sense of laying down of being.

It is in the sharing of this sensibility, and only in this sense, that I see that there is a commonality between very disparate lines of research in various fields, and which has been lumped under the term self-organization, englobing from physics to language and communication.

Second

As for my second closing comment, note that from the previous outlines of how neurobiology and evolutionary thinking is changed by changing our stance, it is evident that such a change does not imply rejection. Again, I see here the possibilities of a rather new style of thinking relative to what has been typical in natural sciences, with a few glorious exceptions.

Consider for example evolutionary thinking and the adaptationist programs. It is true that a change in stance gives a different view of life. However, the selectionist stance need not be negated, and in fact, is the best explanation for the way in which constraints would act. A typical example would be the tremendous environmental changes during the triassic and their impact on the flora and fauna that survived.

Thus, it is possible to achieve a superimposed view of the kinds of systems these approaches generate, provided we keep a *clean epistemological accounting*, i.e. that we keep a clear track of the kind of stances we take in each case. Thus my plea is not for exclusivity of one view, but for a *metalanguage*. In the diagram below I have tried to evoke the transition from one to the other point of view for the case of the study of the brain.

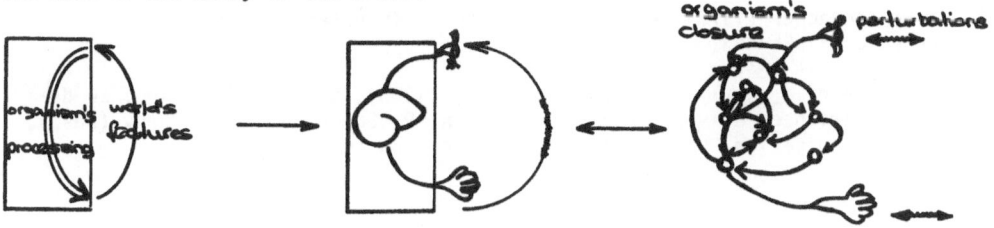

Clearly these two views (input and closure) are not contradictory, but the important point is to recognise that they lead to radically different consequences, and to radically different experimental approaches as well. I do not know whether we might not need other stances which are as distinct and clear as the ones I have outlined here. Maybe we do need them precisely for the study of social systems. This is an open question as far as I can see. But what is clear is that in order to study life and cognition, we do need to explore the almost entirely unexplored land of autonomous-closure machines, clearly distinct from the classical Cartesian-input machines (13).

Notes and References

(1) For this distinction see H. Maturana and F. Varela (1980), Autopoiesis and Cognition, BSPS 42, D. Reidel, Boston.

(2) The notion of operational closure is motivated, defined, and explored extensively in F. Varela (1979), Principles of Biological Autonomy, North-Holland/ Elsevier, New York.

(3) Cf. Varela (1979), op.cit. for further discussion on the meaning and formal definition of this term.

(4) A discussion of recent results is F. Fogelman-Soulie (1984), Frustration and stability in random boolean networks, Discrete Math. (forthcoming).

(5) For more on this notion see H. Maturana and F. Varela (1984), Evolution: natural drift through the conservation of adaptation, J. biological, social Structures (forthcoming).

(6) See Maturana and Varela (1980) op.cit. for the original discussion of this point of view.

(7) For details on this issue, but from a rather different perspective, see E. Land and J. McCann (1971), Lightness and the retinex theory, J. optical Society America 61: 1-11.

(8) For an interesting approach to this question see P. Peretto (1984), Statistical properties of neural networks, Biological Cybernetics (in press).

(9) See Maturana and Varela (1984) op.cit. for more on this crucial point.

(10) See D. Wake, G. Roth, and M. Wake (1983), On the problem of stasis in organismal evolution, J. theoretical Biology 101: 211-224.

(11) S. Gould and R. Lewontin (1979), The spandrets of San Marco and Panglossian paradigm: a critique of the adaptationist paradigm, Proc. Royal Society Series B. 205: 581-598, p. 594.

(12) A.Scheleracher (1972), Divacariate patterns in petecypod shells, Lethaia 5: 325-343.

(13) This paper is partly based on F. Varela, (1983), L'autorganisation: au-delà des appariances et vers le méchanisme, in: P. Dumouchel, J.P. Dupuy (Eds.), L'Autorganisation, Colloque de Cerisy, Eds. du Seuil, Paris.

Can Synergetics Be of Use to Management Theory?

H. Haken

Institut für Theoretische Physik, Universität Stuttgart, Pfaffenwaldring 57/IV
D-7000 Stuttgart 80, Fed. Rep. of Germany

1. Introduction

Synergetics is a rather young field of interdisciplinary research. According to its definition it is concerned with the cooperation of individual parts of a system that produces macroscopic spatial, temporal or functional structures. It deals with deterministic as well as stochastic processes.

This field originated from physics, but over the past years it has found applications to numerous other fields not only in the natural sciences, such as chemistry and biology, but also in the humanities such as sociology. In this paper an attempt will be made to elucidate the question whether the concepts developed by synergetics may be applicable to management theory. The main goal will be the study of processes of self-organization in a sense we shall discuss below.

It may be useful first to describe the spirit of our approach. In physics, chemistry, and biology the self-organized formation of structures is observed and we shall provide the reader with a few typical examples. In the next step of the synergetics enterprise a mathematical treatment is given to these self-organization processes. It turns out that they are governed by specific mathematical relations. But - and this is in the present context the main step - an abstract formulation of these mathematical relations can be given which goes beyond the mathematical formulation based on formulas. We thus find general rules, and we wish to indicate how such general rules may find applications to management theory and related fields. Since we do not expect that the readership of this article is interested in the detailed mathematics we shall make a short cut, namely we shall more or less directly proceed from the examples of physics, chemistry, and biology, to the interpretation of the abstract formulation.

table 1

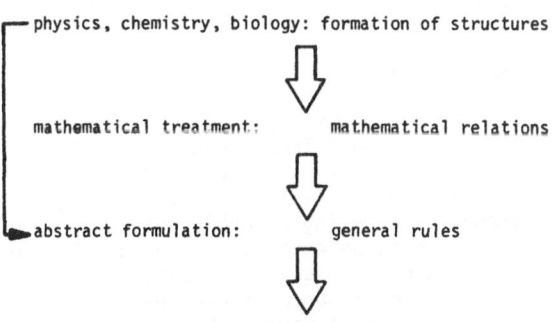

2. Organization versus self-organization

Since in the discussions at the St. Gallen meeting on which this book is based different participants interpret the words "organization" and "self-organization" in different ways, I should like to give a precise definition at the beginning.

In a system various kinds of command structures can be implemented. For instance, according to fig.1 a master computer can give orders to other computers. Or, in a totalitarian country the government gives its orders to the citizens. The reader should be aware of the fact that, for instance, H. von Foerster would even call that latter case "self-organization" because the government has to organize its decisions by itself. I rather would consider the situation as being organized at least as far as the individuals are concerned. But what is really relevant in the present context is the following remark.

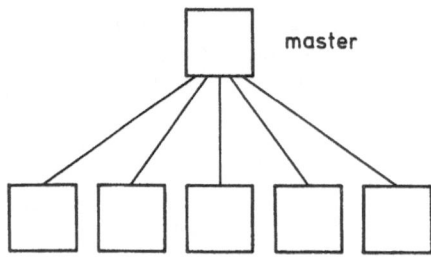

fig. 1: Master computer and satellites (compare text)

Let us consider a system in which no implemented command structure exists between hierarchical levels. Let us further assume that the system consists of individuals (or subsystems) which have all the same properties and which in some way or another interact in the same way. Then we wish to show that even such a system can organize itself into a new macroscopic state with a well-defined structure or, in other words, with a well-defined organization. To illustrate it at the level of management of administration let me give an example. With an implemented command structure we would expect that the boss divides the labor between the individuals of a group and prescribes each individual precisely what he or she has to do. In the process of self-organization on the other hand, we expect that the boss gives just a general frame but leaves it to the group how its individuals will split the labor and cooperate in a maximally efficient manner. I wish to show that even simpler systems in physics, chemistry, and, of course, in biology manage to solve the problems of self-organization in the sense described here. Since the field of synergetics, in particular in the natural sciences, has become very large (cf. the 25 volumes so far published in the Springer Series in Synergetics) I just want to pick out two typical examples taken from physics.

3. The laser paradigm and the fluid dynamics paradigm

A laser (fig. 1a, b) is a new light source which produces light with properties which are basically different from those of conventional lamps. Let us take the example of a gas discharge lamp. In it the individual atoms are excited by means of the electric current. Each excited atom can then emit a light wave track. In this way a light wave is created as if someone is throwing a pebble into water. In such a lamp the excited atoms make their transitions entirely independently from each other, i.e.,in an entirely unorganized fashion. A highly fluctuating light-field results, resembling a water surface into which a bunch of pebbles has been thrown. On the other hand, in a laser the emerging light wave is beautifully regular, having the form of a sinusoidal wave.

34

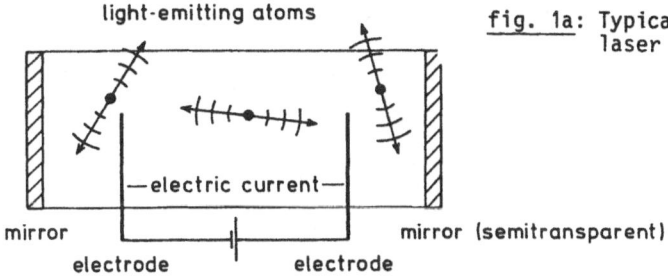

light-emitting atoms

—electric current—

mirror   mirror (semitransparent)

electrode   electrode

<u>fig. 1a</u>: Typical experimental set-up of a gas laser

laser light

<u>fig. 1b</u>: When laser action takes place, laser light is emitted through the semi-transparent mirror

In order to understand the basic difference between the light of lamps and of lasers let us adopt an anthropomorphic picture. Let us visualize the atoms by men standing at the border of a channel filled with water. The case of a lamp would be mimicked by men pushing their bars into the water in an entirely uncorrelated fashion (fig. 2). The regular laser wave on the other hand can be understood in our picture only if the men push their bars in a well regulated periodic fashion into the water. In human life in the latter case we may assume that there is a boss who all the time gives his commands to the men when they must push their bars into the water. But in the laser there is no obvious boss who gives his orders to the atoms. Here we have an example of self-organization in the sense introduced above. However, it will turn not that the system is able to generate its own command structure, namely though the lightfield is generated by the atoms, eventually the lightfield becomes able to regulate the motion of the electrons within the atoms.

The self-organization in the laser has a physical effect which may make some manager envy the laser. At the laser threshold, when self-organization takes place, the efficiency jumps dramatically (fig. 3). Here efficiency is defined as the differential increase of·output versus differential increase of input. To penetrate deeper into the formation of structures let us now consider a second example taken from fluid dynamics.

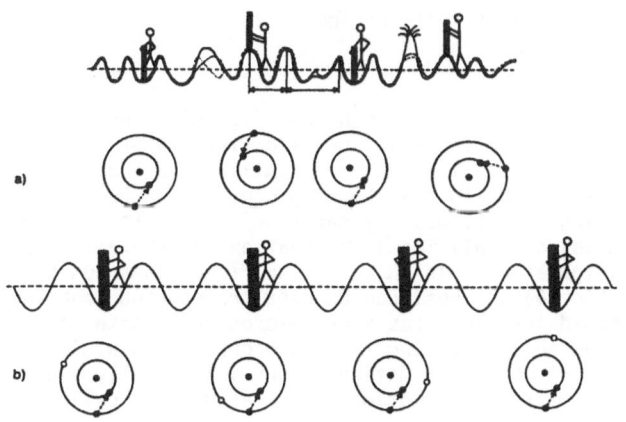

a)

b)

<u>fig. 2</u>: Compare text

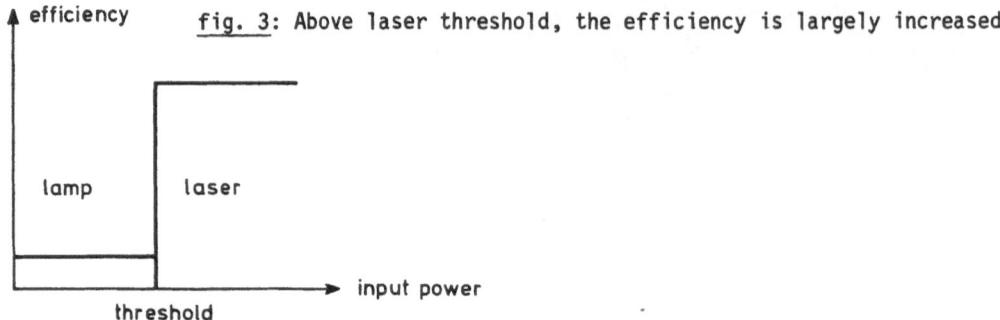

efficiency

fig. 3: Above laser threshold, the efficiency is largely increased

lamp

laser

input power

threshold

We first remind the reader of some basic concepts of mechanics, namely that of
stability and instability (cf. fig. 4). In the upper part we deal with a stable
equilibrium. In this case, after the ball has been displaced from its equilibrium
position, it will return to it. In the lower part, once the ball has been displaced from its equilibrium position, it will never return to it (unstable equilibrium).

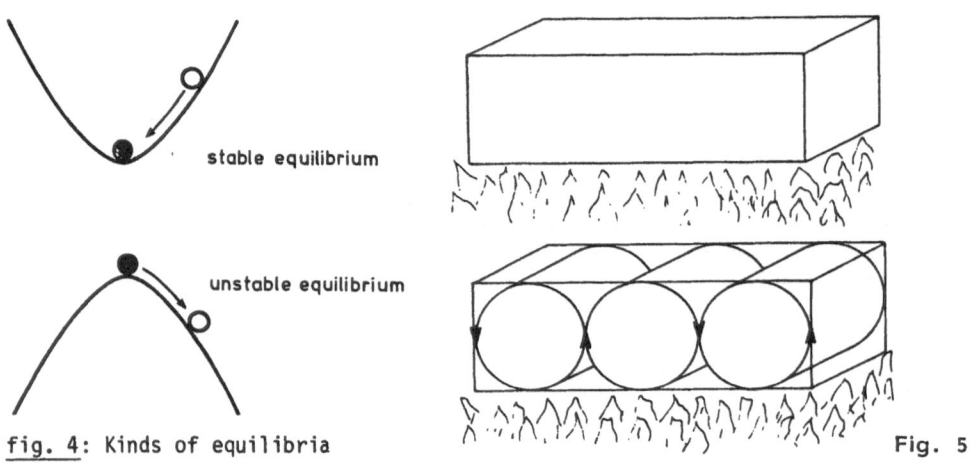

stable equilibrium

unstable equilibrium

fig. 4: Kinds of equilibria

Fig. 5

fig. 5: Pattern formation in fluids.
       upper half: Scheme of experimental set-up.
                A liquid in a vessel is heated from below
       lower half: Spontaneous formation of rolls of the
                motion of the fluid

Now consider a fluid in a vessel which is heated from below (fig. 5). Since the
heated fluid expands, the lower part of the fluid is specifically lighter than its
upper part so that an unstable situation results. This is similar to a crowd of
people where one half of them whishes to get into a shopping center and the other
half wants to get out of the shopping center. Usually pandemonium appears. Nature
proceeds much more cleverly. By means of small local fluctuations of the velocity
of the fluid it tests several configurations. In this way the fluid finds out that
in a specific configuration (cf. fig. 6) the heat can be best transported from the
lower to the upper surface. The fluid then acquires a new macroscopic state of
motion having that ordered form. Now an interesting feature occurs. When we in-
ject ink into the fluid, after a short while the motion of the ink is fully subduced to the macroscopic motion. The macroscopic motion serves as what is called

36

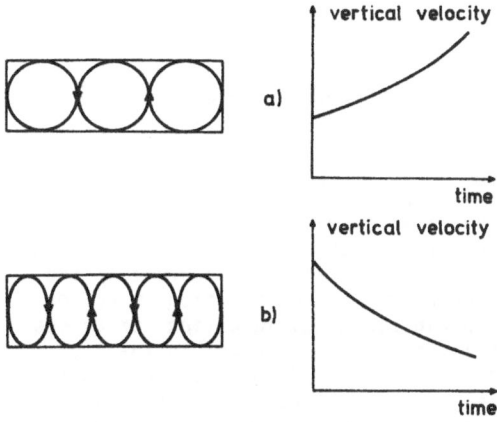

fig. 6: Different macroscopic structures (left hand side) can grow (upper right), or decay (lower right)

a)

vertical velocity

time

b)

vertical velocity

time

in synergetics "order parameter". That established order forces the individual particles to follow the general motion or, to use a terminus technicus, the individual parts of the system are *slaved* by the order parameter. While the motion of the particles, taken as a whole, constitutes the macroscopic pattern, i.e. order parameter, the order parameter reacts on the parts by slaving them. The same holds in the laser case. Though the field is produced by the atoms, the lightfield slaves the atoms in such a fashion that the well-ordered field emerges. As is shown in synergetics, the order parameters and their slaving property can be derived from a rigorous mathematical treatment.

Avoiding all mathematics in this present context, the following slaving principle can be formulated. Let us consider two kinds of quantities which describe the system, e.g.,the order parameters on the one hand and the subsystems on the other hand. Let us assume that these two quantities interact with each other. Let us further assume that the set of order parameters changes but little with time but that the subsystems would change quickly if there would be no coupling to the order parameters. In such a case the order parameters will slave the subsystems, i.e.,the subsystems must adopt such states or motions which are prescribed by the order parameters.

Let us illustrate the slaving principle by an example taken from quite a different field. The language of a nation is a quantity which varies but little over the lifetime of an individual. But when a baby is born it will be subjected to the language of its parents. It learns the language and is thus slaved, to use this terminus technicus, by the language. Eventually, being grown up,it will carry on the language.

I am fully aware of the fact that a number of sociologists deeply dislike the term "slaving" when applied to social context and that they are not satisfied with my explanation that slaving is a terminus technicus which has no ethical or other implication. My opinion is rather contrary and has even changed over the years. In earlier times I strictly stuck to the interpretation as a terminus technicus. I get more and more convinced that in spite of their freedom, humans are much more enslaved than they are usually aware of and that it may be even rather healthy to get aware of that fact. Just to mention again an innocent example. Fashion acts as an order parameter under which individuals are enslaved. In all these cases individuals are sucked into a collective attitude.

We leave it to the reader to invent more examples, some of which become much less innocent when we deal with political opinions,etc., but let us return to problems of management theory.

4. How a new synergetic state evolves

In order to answer the question whether ideas of synergetics are applicable to management, let us first analyze the occurrence of ordered states in physics more closely. Since the system must leave its old state, for instance the liquid at rest, and form a new state, it must be destabilized. In physics this occurs by increase of the pump power in the laser, or by increase of temperature in the case of a fluid. May be this corresponds in human affairs to the collective desire to leave the old state. In the second step fluctuations take place, e.g.,local fluctuations of the velocity field. Translation to human affairs may consist in the following. New ideas must be produced spontaneously, or new inventions made, or new innovative suggestions. In this context brain storming, intuition, or discussions within a group, or with other experts may be mentioned here. In the last step a coherent macroscopic state is formed after a test period based on fluctuations. The formation of the coherent state can be compared to the solution of a jigsaw puzzle. The sometimes conflicting local fluctuations must fit into a macroscopic picture which has certain properties, for instance to maximize heat transport, or to maximize laser light efficiency. This process can be perhaps tentatively translated again into the human domain. A general consensus to establish a new coherent state must be found, possible global structures or processes must be anticipated or presented. Clearly this requires a pronounced feedback between the hierarchical levels of management. Due to this feedback mechanism there will be a continuous change of local structures to match possible global structures.

5. The slaving principle

In section 4 we formulated the slaving principle of synergetics which in a brief version states that long living systems slave short living systems, but it is assumed that both systems interact with each other. To take an example: a stone is a long living system compared to which an individual is short living, but there is no interaction between both so that this principle is irrelevant. On the other hand we mentioned the example of language. Let us supply the reader with a few more examples leaving it to his judgement to agree or disagree.

There are numerous social institutions which in my opinion can be considered as acting as order parameters. Let us consider in more detail the role of individuals played in such institutions such as universities. Since professors stay for a long period at a university, whereas students stay there only for a comparatively short time, I assume that the spirit of the institution is carried on by the professors rather than the students. One interesting question arises when we study the relation between a minister and his ministry. If the ministers change rather often while the officers of the ministry can stay there for their lifetimes, to me the answer in the light of the slaving principle is rather obvious and has been confirmed in talks with a number of officers (may be a minister would give me a different answer). The situation changes entirely if the minister can make his officers "short living" by having a possibility of firing them at his will.

Other types of long living order parameters are provided by local office climates where in most cases a newcomer has no chance of changing the climate. This may have in my opinion interesting consequences for management, namely one has to reckon with the persistence of group attitudes. If they are unwanted, they can be overcome only by dissolving the group or by a job rotation. It may be worth noting that the analysis of synergetic processes in nature and its application to sociology has lead me into the following theory of revolutions. First the total system must be destabilized and then a resolute group of people must take power. Some published analysis of revolutions seem to confirm this conclusion.

6. Why is self-organization desirable?

The problem of a structure, which in our context is called "organized", consists in the information bottle neck. Here all information of the performance of the sub-systems must be provided to the central unit where it is processed and eventually new orders are given to the subsystems. In many cases this procedure may take too long a time so that the system cannot perform as wanted. This problem can be solved only if more room is given to self-organization at the lower hierarchical level and only the really relevant features are communicated between the upper and lower levels. In the next section we shall see how even large self-organizing systems can be rather easily steered when basic rules of synergetics are observed.

In conclusion of this section I mention that in physical or chemical processes inadequate external steering of otherwise self-organizing systems can lead to chaotic, i.e.,entirely irregular processes or, in other words, to steer a synergetic system requires a knowledge of the general laws of synergetics.

In the present article I cannot dwell on these typical features but I wish to give a specific example of far-reaching consequences.

7. Conflict situations

When physical or chemical systems are driven to an instability point, beyond that point they can acquire new macroscopic states. But, and this is a very important feature of these processes, in many cases the newly evolving state is not determined uniquely a priori. For instance,in the case of fluid rolls the central roll can move to the right side as well as to the left side. The situation is symmetric with respect to left and right motion, but when the motion is realized, only one possibility can be chosen and in this way the symmetry is broken (fig. 7). Symmetry breaking can be used as a paradigm for many problems in human life, for instance decision making. Managers are occasionally blocked in their decision by seeing two different possibilities for the solution of a problem both of which have their advantages and disadvantages, and they waste a good deal of their time invoking more and more criteria. In my opinion, when after a certain time no criterion suffices to make a decision, then it does not matter which decision is actually taken. At that stage of knowledge both are indeed equivalent. So decision making means to break symmetries in a way arbitrarily.

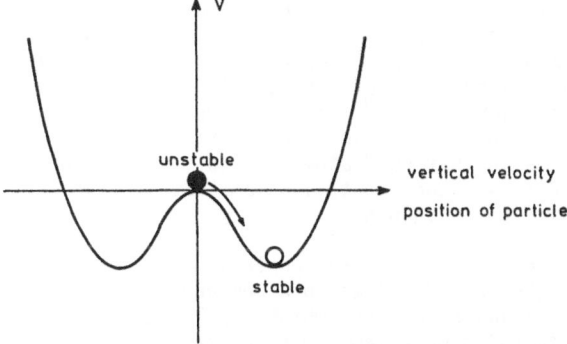

Fig. 7: Visualization of the transition of a liquid heated from below from its unstable to a stable state. The vertical velocity of the liquid (at a chosen position) is formally identified with the position of a particle moving in a vase

In a number of cases such as in fluid motion, it has no deep consequences whether the fluid rotates in one or the other direction. In other cases quite different situations are concerned which may be visualized by fig. 8 presenting an example from the field of perception. In self-organizing systems a small fluctuation may decide whether one or another macroscopic state is adopted. When in management we let a group of coworkers choose their own mode of operation we may expect various modes which may depend on some initial accidental events. Such collective behavior may be collective activity in achieving a goal, or,e.g.,collective laziness. This strongly indicates that at least in the beginning of a process, which should lead to self-organization, external symmetry breaking must be achieved by giving the initial state an asymmetry. For instance, to steer a group of people in the beginning quite actively, and then relax this steering, or having initially a number of particularly active people in a group.

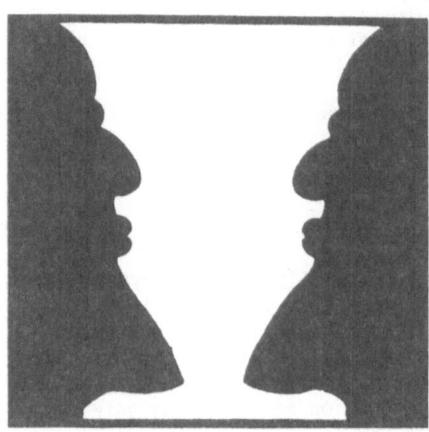

fig. 8: Example for symmetry breaking.
Vase or faces?

Let us finally discuss the question   how far a command structure must be implemented into a system. From what we have seen above it follows that systems obeying certain mathematical rules (which are definitely not physical rules) can organize themselves without any implemented command structure. In such a case new collective motions or attitudes evolve which are described by order parameters. In other words, the systems can create their own structure or organization. But the reader must be warned to develop an unlimited enthusiasm about self-organization. The explicit examples of physics show that quite often it takes a rather long time until the new ordered state is formed, i.e.,it might require too long a time to be established. Furthermore there is only a finite amount of resources available in life. Finally and most important it cannot be excluded from the outset that self-organization leads to undesired modes of behavior (e.g.,collective laziness,etc.). This leads to the conclusion that some command structure must be implemented. But according to the results of synergetics, in general such a command structure is not of the direct type in that each atom, for instance, is prescribed from the outside how to move. Rather global control parameters such as the energy input into a physical system have to be changed.

When we try to translate this concept into human affairs, control parameters would be represented by taxes in economy, or by rewards given in a company,etc. Symmetry can be broken by setting appropriate initial conditions which can be achieved in a company by hiring excellent people to build up a new group. At the same time synergetic systems may easily adjust to new situations because by means of fluctuations they can respond to such changes. As a consequence sufficiently large internal fluctuations, e.g.,making new suggestions or to test new organizational forms,should be allowed in management.

40

8. What is synergy?

Since the words "synergetics" and "synergy" sound similarly, one may expect that synergetics gives an answer to what synergy is. The concept of synergy is a rather old one in management theory. But in this author's opinion it is not well defined, because it seems that it is understood as a static property. What counts in the context of synergetics are processes and their efficiency, i.e.,in our understanding "synergy" is a dynamic property of a system. Roughly speaking we should say that a system has a high synergy if its processes are well correlated producing little internal friction, i.e.,that the processes are coherent so that the individual processes fit well together,yielding a high efficieny of the total output.

9. Concluding remarks

In my article I have tried to elaborate on some basic concepts of synergetics and its possible applications to management theory. Readers interested in the mathematical apparatus are referred to my books "Synergetics. An Introduction" and "Advanced Synergetics". A popularization of synergetics with applications to numerous disciplines can be found in my book "Erfolgsgeheimnisse der Natur" (DVA 1980). The English translation appeared in 1984. ("The Science of Structure: Synergetics", Van Nostrand Reinhold, New York.) The reader should be fully aware of the fact that synergetics is not intended to replace management theory, but rather we hope that some of its concepts may prove useful for that theory, possibly along the lines briefly indicated above. In my opinion the situation may be similar to that of bionics. As is well known, this branch of engineering studies biological objects, e.g.,the shafts of grass,in order to find out in which way to construct a television tower. Similarly, in synergetics we explore how nature (in particular biology, which we did not treat here explicitly) manages to form its organisms by means of self-organization. May be along this line still more can be learned how to handle problems of management.

References

Haken, H.,	Synergetics. An Introduction, Nonequilibrium Phase Transitions and Self-Organization in Physics, Chemistry and Biology, 3. edition, Springer Verlag, Berlin, Heidelberg, New York, Tokyo 1983
Haken, H.,	Advanced Synergetics, Instability Hierarchies of Self-Organizing Systems and Devices, Springer Verlag, Berlin, Heidelberg, New York, Tokyo 1983
Haken, H.,	Erfolgsgeheimnisse der Natur. Synergetik: Die Lehre von Zusammenwirken. DVA Stuttgart 1981 English translation: The Science of Structure: Synergetics, Van Nostrand Reinhold Company, New York 1984

Also: Springer Series in Synergetics, Vols. 1 - 24, Springer Verlag, Berlin, Heidelberg, New York, Tokyo

Self-Organisation: Some Theoretical Cross-Connections

R. Riedl

Institut für Zoologie der Universität Wien, Althanstraße 4, A-1090 Wien, Austria

When, as in a cartoon film, we observe building bricks rising of their own accord, turning and arranging themselves into a wall, a gate-house or a castle, we speak of self-organisation. We expect that the castle was indeed one of the potentialities present in the box of bricks, but that there were so many potentialities that it was impossible to foresee them. The building bricks were pre-disposed.

However, we do not expect that they were pre-determined, let alone predestined. That is, there were no prior forces or impulses, nor goals or purposes, that might have let us foresee the rise of this particular castle, which would simply have escaped our foresight. This is odd, for we can indeed ask what causes or purposes were at work in completing the construction, although they were not at first available. The key term is "at first", for as we proceed things should go ahead in a normal manner, with physical forces and natural conditions of selection. Still, we expect that at any stage there would be several motives and selective possibilities and it would in large measure be left to the chance of prior conditions which possibilities will necessarily be chosen. This involves genuine micro-physical chance, which in running through long causal chains is known to reach even into macro-regions. We thus exclude from our concept of self-organisation those differentiating processes that, because of their antecedent conditions and programmes, move towards a foreseeable goal. These are teleonomic processes; for example, the genome and environment of a healthy fertilised chicken egg allow us to expect that a chicken will hatch, a rat will in the end traverse the most complicated maze, we can learn to master the integral calculus.

What we have here are processes of phylogeny, not ontology; not a renewed formation of order, but the formation of new order that was not present in the prior conditions even vestigially. This delimits our object.

On corresponding phenomena

Self-organisation occurs at all levels of our complex world, from physical systems to cultures and their artifacts. In this we observe a hierarchical structure of these systems always involving several lower systems in the building of the next higher. Thus quanta make up atoms, these in turn molecules, bio-molecules, ultra-structures, cells, tissue, organs and organisms, and these the layers of societies and cultures (Riedl 1976). The development of this layered hierarchical structure proceeds along the time axis of cosmic, chemical, biological, social and cultural evolution at growing acceleration. Along this development new regularities like-wise arise, and, what is the same, the possibility, in principle, of foreseeing states of order and organisation. The degrees of order in systems or the extent of attainable foresight may be formulated as "law times application", a determined correlation multiplied by its redundant repetitions in the world's objects. As systems evolve, starting with 10^{80} identical representations of the quantum laws, we find that the amount of redundancy falls; or conversely, the amount of order or differentiation grows.

Example: 10^7 bricks can be stored in stacks of 100 by 50 by 20 in a complete but highly redundant order. The same tiles in a brick dome have the same order but with much less redundancy. The amount of order or differentiation has risen in line with the co-ordinates for the position of each individual brick. Nevertheless a chaos of 10^7 bricks tipped over a slope would require the greatest amount of description, being quite devoid of redundancy.

According to our way of seeing things we tend to consider degrees of order only in levels. The order of lower levels is taken for granted, in the sense in which we take any structure as in turn consisting of substructures.

Example: the statement "the children's room is a chaos" means that the toys have reached a chance distribution, while the toy clock remains intact. "The toy clock is broken as well" means that its parts have lost their functional connection, while individual gear wheels remain intact. And so on.

Degrees of order and content of redundancy, amount of order and differentiation can in each case hold for the systems at all levels, but always in that hierarchical relation in which we consider them a level at a time; just as we speak of different laws at each level, for example those of chemical bonds, breathing, social ranks, or Gothic art. A further relation that seems to hold for all levels of self-organisation concerns the kind of, and direction in which, causes appear in the rise of new systems: for the latter are not simply superstructures, mere syntheses from what already exists, but rather insertions between prior parts and an equally prior whole. This matters, because it points to two sets of causes, starting from higher system and subsystems respectively, and differing in kind.

Example: galaxies arise between the parts of material clouds and the gravitational field of the whole cosmos; proteins between concentrations of molecules and the total surrounding of the planet's secondary atmosphere; tissues between cells and the individuals of the whole population; professions between the needs of individuals and those of a whole culture.

The effects from subsystems are described by two sorts of causes: those of the forces or impulses issuing from the subsystems, or efficient causes; and those of the available material parts and the way they can be put together, or material causes, in the sense of Aristotle. The effects from the higher system again are of two sorts, both concerning conditions of selection. They likewise determine which and how many parts are linked and ordered as promoting the conditions of maintenance for the new system. Only those attempted new systems maintain themselves that satisfy the conditions. The static or material view of such results of selection we experience as form, organisation, or plan (in the architectural sense), corresponding to formal causes. The dynamic or functional view is what we call functional purpose or point of a thing, or final causes, again in Aristotle's sense (Riedl 1980).

With regard to the principal axis of the stratified hierarchy of the world, there is a double symmetry: efficient and material causes act from subsystems, formal and final causes from hypersystems. Severing finality from causality, as has been usual from the first misinterpretations of Aristotle right up to the present, is inadmissible: like biologists today, Aristotle understood finality not as teleology, but as programmatic teleonomy (W. Kullmann 1979, 1980). Likewise for physicochemical, genetic and consciously intended programmes.

This symmetry arises from the mirrorring of that cognitive dualism which makes us look at state and action of one and the same object in two ways: it divides the world into strata of body versus soul, matter versus spirit, and over the whole of organismic structures versus functions up to the seeming duality of matter in its manifestations as wave versus corpuscle, and ultimately energy versus information, where the selective conditions of the hypersystem through their exclusions give rise to information that grows in its content of order and effort at description, construction and decision. The second symmetry consists in the material side of causal relation, namely material and formal causes, appear as changed conditions for each layer. That is why we have differentiated these conditions into the terminology of the different sciences, of physics and chemistry, molecular biology, cytology, histology and so forth up to psychology, sociology and the humanities. By contrast, concepts of force and purpose, efficient and final causes, are taken as reaching through the whole stratified system as far as we think we can compare them with our forces and purposes. As regards purposes, this is only as far as organic relations, while forces are taken to be in evidence as identical right down to quanta and right up to a power block.

The operative paradigm is that the dynamic and material regularities of all lower strata penetrate all higher levels. How mutual effects of upper layers influence lower ones has been considered as of less interest.

Example: we presuppose the same electron shells in atoms and in bio-molecules, and these in breathing and that as pre-condition of group dynamics which figures in the laws of change of scientific paradigms. However, we tend to remain unaware of how the choice of a paradigm influences the dynamics of participating groups, or stress and satisfaction the breathing of individuals, breathing frequency the activity of participating molecules. Only in the domain of intention are we aware that the plan of a building determines its rooms, these the walls, bricks and their combination.

If we recognise shared properties in the various layers of self-organisation, such as degrees of order, differentiation, duality of both grounds of genesis and grounds of explanation, and the pentration of regularities, we may ask whether consonant conditions of self-organisation are likewise to be expected.

Some shared explanatory terms

Those who have written on self-organisation were doubtless mostly biologists, and next physicists and economists, witness the works of Ashby, v. Bertalanffy, Eigen, v. Foerster, Haken, Jantsch, Lorenz, Malik, Maturana, Prigorine, Probst, Riedl, Simon, Ulrich, Waddington and Weiss. Some of us wrote their piece with the aim to explain the process of self-organisation in general terms. However, we can see today that it often concerned only one of the terms that contribute to explanation, while leaving others out of account. Let me here give a comparative summary of what may be the conditions that hold in all strata of complexity, as if ambitiously aiming to help prepare a metatheory of the self-organising process. To limit the task and make it the more memorable, I shall speak only of three phenomena (as in a rondo).

First consider the three preconditions (1-3) in a presupposed regular universe with highly redundant phenomena and open systems describable in terms of the change of energy, entropy and information. The indispensable conditions are

1. *Variation* of the environment in which the system develops: physically these are breaks in symmetry, chemically concentrations, biologically conditions for biological niches, sociologically conditions of trends and culturally conditions of ideas; always in a non-Euclidean, non-Cartesian, Aristotelian space.

2. *Fluctuation* in the considered systems themselves, the ability to loosen rigidity or determining conditions. Physically these are instabilities of energy,

chemically gradients in the kinetics of reactions, genetically drifts, anatomically adaptive radiation, socially change of paradigm and revolution.

3. *Competition*, which presupposes conflict in a plurality of comparable systems, with the object of conditions for maintenance; and a *selection* that dissolves systems of smaller correspondence. A selection by hypersystems as breaks in symmetry, selection of surroundings and market or social self-evidences lead to adaptations; a system selection as molecular parliament, as internal operating selection or power of judgement leads via organisations, management or reason to higher speeds of adaptation. (System selection is so little known that we shall return to it later).

Alongside these preconditions for any process of self-organisation, we note three basic conditions (4-6) that must be given in the systems themselves.

4. *Stabilisation* as the ability to swing back from medium disturbances into stable states. The phenomenon is known physico-chemically as equilibration, but biologically rather as homoiostasis or regulation and in psychology as flexibility. It stands between the extremes of rigidity and disequilibrium or instability.

5. *Innovation* as the change of self-modification into basically new states, to which the participating parts of the system are predisposed but neither predetermined nor predestined. According to time, place or type of formation they are accidental. Physically we speak of new motions or structures (Benard cells, shore ripples), chemically of new properties, biologically of mutants or fulguration, socially of cultural mutants or ideas; where, in complex systems, the rate of innovation must match the content of information.

6. *Preservation of chance*, that is, of genuine micro-physical chance (not the cognitive kind of mere ignorance) right up into macro-physical systems; either through linking macro-physical determinants with micro-physical chance events, as in the mutability of biological systems, or, as often occurs, through complexity, namely long causal chains, again from the parliament of molecules to creative ideas. For of these last none is wholly contained in its premises, or else we might make all possible remaining discoveries and inventions today.

Only when these six conditions have been fulfilled can those phenomena appear that we give the honorific title of "evolutionary". We mention only three of them (7-9).

7. *Hierarchical differentiation*, in the sense of transformation of complexity into complex order. In other terms: in the sense of dismantling redundancy and building up of sub-ordinate interdependences, or new regularity, as we call it. This principle first succeeds under economic or competitive conditions because of greater homoiostasis as against stochastic disturbances. Later, in extended function, it promotes the speed of adaptation and the storing and recall of data. In all layers it is called differentiation or organisation.

8. *Reproduction and memory* become evolutionarily successful because in a regular though variable and redundant cosmos a plurality of the same system must enhance for its principle the chance of success, adaptation, or in short, preservation. With reproduction it becomes necessary to transmit and reproduce information. Thus in inorganic science we speak of allosteric auto-catalysis, in biology of heredity and in culture of tradition, where each of these layers is a precondition of the next.

9. *Creativity* in the sense of a guidance or channeling of innovation, by conservation of hitherto successful interdependences within the system. In other terms: by exclusion of nonsense, or of the proved, from the search-field of innovative chance.

This moreover keeps the search-field small, for the chance of a successful hit is the inverse of the number of lots. From this follows the adaptability even of complex things, and the autonomy of the system with radiation of its degrees of freedom and the channeling of that adaptability. Layer by layer we speak of the given codes, the epigenetic system, the archigenotype (Waddington 1957), homology, morphological type, instincts, innate forms of intuition, convictions, social and cultural attainments.

In the field of evolution, the field of self-organising processes touches on the evolution of evolutionary mechanisms. This still involves the reaching through of the deeper laws of a layer, but also superstructure that can have no vestiges in the lower layers. So much for what is common to layers.

On the self-organisation of reason

Reason, as is well known, cannot be based on itself alone. The trilemma of cognition, according to Albert (1968), admits only recognition of circularity, an infinite regress, or cutting off the process. Even the reality of the world cannot be proved by reason. A distillate of all a priori categories of reason, which can thus not question them, are the postulates of space and time, and chance, comparability, causality and finality, as a priori in the sense of Kant. This raises the philosophical problem of isomorphism, for how do a priori features of reason fit into this world?

The "evolutionary theory of cognitive gain" can now answer this: we recognise the a priori and hereditary forms of intuition of reflecting reason as a posteriori products of learning through reflection: the a priori forms of intuition arose before any reflection, as products of self-organisation, and thus part of our present subject.

We consider evolution as a process of cognitive gain, for we can show that success in life depends on extracting the regularities of the relevant environment in each case, as an aid in deciding that ensures successful forecasts. Thus the learning of genes has extracted from our surroundings all the optical laws at a certain level of differentiation, incorporating these in the structure and working instructions of the eye, to such an extent that physicists, using precisely the optics already built into them, have merely rediscovered the laws of optics.

It is therefore appropriate to speak of creative learning. If we wish to reserve the concept "cognition" for cognitive processes, it is always a question of cognitive gains, because the concept of "informational gain" has nothing to do with correctness, let alone with relevance to the need of receiving information. Moreover, we rightly speak of extraction, since knowledge can be gained only in so far as the environment contains regularity accessible to the extractor.

Three things that follow from this theory are important to the process of self-organisation: first, the penetration of self-organisation right to the conditions of our intuition; secondly, the limits of those products of self-organisation, which help us to intuit self-organisation and therefore also the relation of evolutionism to constructivism; thirdly, the possibility of this form of intuition transcending itself.

As to the first: our forms of intuition turn out to represent the world's regularities, but in a highly elementary way, for the mastering of vital tasks in the as yet very simple environments of our distant ancestors. Spatial and temporal sense are probably as old as vertebrates, the most recent form of intuition (that of purpose) is older than the genus Homo, about 400 and 5 million years respectively.

They therefore do not picture the world and certainly not the "being-so" of things; rather, we must take them as aids for deciding that must be valid for very general tasks in the environment of long chains of generations. They are narrow passages in sense, or rather interpretation, into the basic forms of order in the world round us, as if in a compass-like arrangement that does not set up internal links. Hence dualisms in the intuition of a world that is not divided in any of the cases, cognitive dualities. Time becomes imaginable as one-dimensional, but space independently as three-dimensional and Euclidean. For our terrestrial microcosm, these are quite adequate sectors from the mutual relation of the four-dimensional space-time continuum, a theory that seems to represent a better approximation to reality.

As to probability, namely a severing of chance from regularity or from necessity or intention, as is of life-preserving importance to predictability, we have the a priori expectation that if our forecasts are confirmed the occurrence of consequential forecasts become ever more probable. This principle guides all associative cognitive gain from conditioned reflex to the assessment of our observations and experiments, or else we should not conduct research. It is the basis of induction, an extra-logical principle, since it contradicts the impossibility of truth-extending conclusive inference. We even incline to the view that if an event has been long delayed (the clouds will disperse) it will become more probable that it will happen.

These two expectations are isomorphic with our world in so far as most repeated (successive) coincidences are not adventitious or because most sequences of events are clustered and depend not on chance alone. Of course, such expectations are often disappointed, but nevertheless correspond to the best amongst simple approximations. This hereditary programme became the basis of empiricism, overlooking that the very programme of any individual experience must be a priori.

As against the empiricist principle there is a projectivist one with programmes of comparison and the perception of shape. We possess this a priori with the expectation that the unlike is to be interpreted by the abstracted like, with missing parts supplied by thought, sequences brought into connection and shapes viewed as spatial (in perspective) and transposable as wholes. These expectations, too, are logical but cannot be established. Isomorphism with the world rests on the latter's redundant order and on the fact that most of its structures are indeed spatial, showing themselves differently according to perspective, distance and concealment, without sudden changes and following certain basic laws. Thus is rests on the mostly overlooked arbitrary freedom of combination of features (now those simultaneous coincidences). In this sense of prior expectation of an ordered world the hereditary programme is the basis of rationalism, which overlooks that the reflecting individual's a priori expectation is an empirically acquired product of genetic experience.

Finally, as we saw, of two of the four types of causes we have for each a uniform intuition, namely of forces and purposes. However, mutual relation between causality and finality is again excluded from intuition.

As regards causality, we have the expectation that like things or results have the same cause, that each thing has its prior cause and that in the chain of these there is an original or prime cause to be discovered, a first impulse of force from which all further movement follows executively.

The isomorphism of this expectation with the world is again an approximation resting on the fact that most sequences of states and events indeed are not merely adventitious, and that this holds likewise for prior states and prior events. We further admit that this is connected with transformation of energy. However, we lack the intuition for causes not being chains, but rather networks, so that their effects do not remain canalisable. The programme is the basis for the theory of materialism, which overlooks the fact that from forces alone no historical system in our world can be understood.

As against the expectation of causality there is that of finality, that like structures or processes have the same purpose, that they are functions of the same hyper-function, that every purpose has its place in a superordinated purpose and that it must be possible to find in the chain of purposes one last or final purpose of all things, such as the intention of their creator.

In this a priori condition of reason (the last, according to Kant), there is a further approximate programme. No doubt, in spite of all possible disappointments, the most often confirmable hypothesis is that like things have the same purpose. However, the consequential expectations of chains of purposes are again most misleading simplifications. In this sense the expectation of a world order pointing to ultimate purposes is the basis for idealism, which overlooks the fact that ultimate purpose can never be established but only postulated and any function can arise only in its material system. Prior to any reflection, the process of self-organisation thus manufactures the preconditions of our conscious reflection in the form of aids to decision or judgements, as prejudgements. These are doubtless fashioned on the pattern of laws and environmental order, but are only general approximations adequate for the survival in an as yet very simple setting. The isomorphism is approximate.

There are indeed signs of an isomorphism of a higher order in which the symmetries of space and time, successive and simultaneous coincidences, forces and purposes, recur in our expectation as well, although of course we have no forms of intuition as to their relation to each other, and this lack has misled us into viewing them as opposites.

On the limits of such self-organisation

As to the second: the limits of such self-organising processes lie primarily in the limits of the environment: to problems it does not contain there can be no reaction nor a solution found. Conversely there are of course life problems that must be circumvented because the system can produce no solution: gills for dolphins, feathers for bats and wings for Icarus.

Obviously all these approximate solutions of life tasks are solutions amongst others. Such alternative solutions are called analogies. We may think of the very different forms of solutions for sight and flight in bees and birds. It is precisely the differences between analogous solutions that show us how distant they are from a maximal or definitive solution. Nevertheless the process is one of optimisation, in which the asymptotically attainable ceiling depends on the possibilities in the system concerned. These are as different for dragon-fly and swift as for helicopter and glider.

Nevertheless there must be a relative agreement with the laws of the real world, so long as this is effective through selection. Still, Watzlawick is right in saying that a captain who passed through unknown straits at night without running aground so far has no detailed knowledge of the coastline, but at least he knows that it did not lie on his course.

This relative isomorphism, however, vanishes if the conditions of selection become indeterminate or non-existent. This may easily occur with the rise of consciousness, because it permits experimenting in imagined space and therefore shifts the testing of hypotheses or expectations from the environment into the organism itself, so that what is thinkable may be confused with what is. More crassly still, the conviction carried by what is imagined has always seemed more real than reality than did "the clamorous crowd of the senses" (Parmenides). If genetically learnt forecasts can develop only relative nonsense, "to believe pure nonsense is a privilege of man" (Lorenz 1977).

Artificial realities however unreal, as found in constructivism,can of course persist only on two conditions: either they are free from any empirical selectivity, or subject to the pseudo-selectivity of so-called social truths . These are just as much constructions with the one difference that they are developed or adopted by a whole group (Berger/Luckmann 1966). We can indeed err, "often for centuries" (Dessauer 1958), until experience one day refutes us after all.

Whether we can think of the world as filled with aether, or with gravitons which so far remain just as undemonstrated, as soon as by much risk-taking and calculation we succeed in putting a man on the moon and bringing him safely back to his family, there must be a certain agreement between expectation and reality. Merely our being here and talking about mistaken constructions can be interpreted only in evolutionary terms. Even our own existence is comprehensible to us only from a certain agreement between the "hypotheses that our organs represent" (Popper 1972) and a reality, whatever "being so" may hide behind our perception. Evolutionary theory and its "hypothetical realism" (Campbell 1966) are the preconditions for existence and criticism by constructivism, for evolution itself is an interplay of invention and criticism.

Thus from an analysis of the limits to our forms of intuition we may gather some foresight of what are the preconscious ratiomorphic deficiencies and rational mistakes of our rationality. For the sum of our hereditarily guided prejudgements about the world we have the concept of "ratiomorphic apparatus" (Brunswick 1957). Its performance corresponds to sound and unreflective common sense. Its judgements and prefabricated intuitions cannot be totally wrong, nor complete. They contain truth, but not only truth, let alone the whole truth.

That is why for a whole set of phenomena that through empirical evolutionary and self-organising processes reveal themselves as necessary assumptions, we have no forms of intuition, though we are well advised not to reject their possible reality on that account.

A by now classical example is our inability to visualise a four-dimensional space-time continuum, so that we can imagine neither the beginning of time nor the end of space. Less well known is our inability to foresee the rise of new qualities from quantitative changes. Concerning the question "how many grains make a heap?" (Hassenstein 1979), we admit after the event that grains roll while heaps slide or flow, but the perception of transition between them creates difficulties for our imagination and leads us to regard the analysis as pointless. This becomes clearer still in complex domains where it is difficult to expect the merging of systems to give rise to new systemic properties that could have no vestiges in the constituents.

Our ordinary language indeed lacks the concept for such "appearances". "Creation" and "evolution" are metaphors of bringing out or unrolling of what is preformed, as in the old doctrine of preformation. We must bury the hope of linguistic neopositivists (Wittgenstein), that in our concepts there are a priori certainties. In fact our language is not even suitable for the concept of change of shape. It is full of misleading analogies, but for example lacks concepts for swimming bladder-lung, primary jaw-auditory ossicles, hair tufts-rhinal horn (of the rhinoceros), although the second in each pair developed from the first (Remane 1971).

What hinders us above all is that we have no inborn form of intuition for the mutual relation that must exist between causality and finality; nor do we have a connected form of intuition for material and plan, disposition and selection, energy and information. Even structures and functions are separated into nouns and verbs in all our languages, although no leg ever developed without walking and nobody would have walked without legs. If in the case of our ratiomorphic performance there are at worst deficiencies, it is noticeably our rational performance that makes serious mistakes, generally because ratiomorphic guidance is taken to agree with reality and we expect to be entitled to extrapolate from it arbitrarily. Yet we know that our forms of intuition represent only approximations so that arbitrary extrapolation must lead to increasing error.

If the three dimensions of space had turned out to be independent of their time and content, as our inborn intuition makes us believe, then the concepts of Euclid and Descartes would apply, and the parallel axiom as well as the world viewed as consisting of separately travelling particles would be binding. Since this expectation turns out to be wrong, all its consequences must be questioned and the spatial concepts of Aristotle and Einstein are the next best approximation.

Had the confirmation of forecasts led to truth even without prior knowledge, we could make do with the empiricism of the positivists. Had our imagination or ideas of shape led to true picturings of the world even without prior experience, then the concepts of rationalism would be sufficient. Since, however, all empiricism turns out to need prior knowledge and all imagination rests on prior experience, all consequences of empiricism and of rationalism are questionable. The separation will thus be false. We can foresee that the same will hold for any separation of causal from finalist solution. Of course we can, as materialist reductionism does, go on dismantling systems and in the end explain them from their impulsive forces, and nobody overlooks the extraordinary success of practical reductionism. However, to say that in this way the whole system is grasped, as ontological reductionism maintains, is grossly misleading. Think for example of the grounds on which a system arises. Idealist reductionism is equally misleading when it states that the systems in this world can be understood from ultimate purposes.

We are indeed endowed with perception of forces and purposes, although without links between them. Our reflecting reason then constructs from the difference the either-or of mutually exclusive solutions.

This circumstance, of having to operate at (or already beyond) the limit of our forms of intuition I see as the main obstacle to the development of concepts of evolution and self-organisation, now and in the future. We need merely consider that along these oppositions our whole culture was split in two (Snow 1959); that people still believe that formal logic contains some certainties, although this was put in question by Boltzmann long ago. The necessary systemic assumptions alone have summoned up a host of enemies (Churchman 1979).

On self-transcendence

As to the third: this leads to the last question on the self-organisation of reason: could we transcend the limits of our hereditary forms of intuition? Lorenz calls them unteachable. However, we ourselves might turn out to be teachable, but under whose guidance? Only that of experience and the failure of forecasts, so long as they do not leave our intuition. Let us consider the case of Einstein.

To me as an evolutionist, Einstein's great achievement consists in his having given in to experience in a conflict between what can be intuited and what is a totally unintuitive theoretical solution to former riddles of experience. His discovery certainly did not change his own hereditary intuition of space and time. The theory has transcended intuition and even added a new dimension.

That is in origin precisely what philosophers understand by transcendence and transcend, the climbing over; starting from the expectation that there must be something that transcends the spatio-temporal world of human intuition. The difference in points of view is indeed considerable: we expect that experience will confirm or refute the products of speculative reason; we expect a natural science of transcendence.

This expectation corresponds to our well known mechanism of cognitive gain in the evolutionary process. The transcendence of our hereditary equipment thus lies directly on the path of our possible development.

This no longer concerns genetic adaptation, which we have no chance to expect. Besides, there is fortunately no selective pressure on adapting our forms of intuition. Our chance lies only in transcending them rationally, since they are unteachable in spite of their suggestive force.

In the case of the theory of relativity some seemingly modest phenomena confirm that the concept is correct. The easiest to ascertain is the deflection of light in the sun's gravitational field (by 2.2 seconds of arc at the sun's edge). This leads to the most incredible consequences. If we travelled at speeds approximating that of light, we should find our surroundings in the direction of travel shrunk. Conversely, a stationary observer would not trust his senses: he would find us totally flattened. If only we could see far and fast enough, wherever we looked we should see the back of our own head. So much to illustrate the unimaginable.

Of course, we should have to travel nearly at the speed of light in order to perceive the error of our inborn forms of intuition. In the microcosmos region of our planet our forms of intuition are adequate approximations, as mentioned.

However, the case seems to stand differently as regards phenomena of probability, comparability, causality and finality. They act without dimensions and therefore even in the domain of everyday life on earth. Here any growth of even modest adaptive deficiencies must become our very own terrestrial torment. This problem is thus much more "relevant", to use current jargon. Only the metric formulation of our errors remains to be given, but that is much more difficult.

However, we have reliable data on the fact that we always approach complex phenomena with models that are too simple, so that we tend mentally to simplify causal connections and look in the skein of mutual dependences for a main or original cause (Dörner 1975, Riedl 1982). More serious still is the error of regarding the causal and the final as incompatible quantities (Spaemann and Löw 1981), and our inability to grasp the mutual relation between energy and information. All this is as old as the doctrine of Plato, almost as old as our civilisation.

Instead of pursuing this matter in general terms, I shall give two examples for illustration. Both span a development in the intellectual history of the last two centuries, and are to show how a manysided causal relation in self-organising processes was gradually rationalised out of existence because it does not match our intuition. One case concerns evolutionary theory, the other epistemology; and both the knowledge of our self-organisation for the sake of cognising self-organisation.

Loss of insight into genetic knowledge gain

We recall that the evolutionary theory of organisms began with the notion of a mutual causality. When creationism, as still maintained by Cuvier, was supplanted by evolutionism, prepared by Maupertuis, Lamarck and Erasmus Darwin, a two-sided causal relation was entertained. On the one hand people presupposed what every animal breeder could observe, that hereditary material determines body structures. On the other hand a mode of operation was sought which conversely explained how body structure influenced the hereditary material (how phenes influence genes, as we put it now). For that such an influence must be assumed seemed beyond question.

Lamarck expected a "feedback message" of changes in phenes by their use and Charles Darwin, sharing this view, developed in his theory of pangenesis (1868) a model for how such a procedure might be visualised. It was the phenomena of regeneration, of taxonomic units and their characteristics, that made Darwin cleave to the presupposition of a mutual causal relation in order to explain evolution.

"Darwinism" did not arise until Alfred Russell Wallace published a book of that title in 1889, seven years after Darwin's death, leaving out the concept of pangenesis and therefore the Lamarck-Darwin notion of mutual causality. Wallace, a pleasant man but less informed, had discovered the principle of selection at the same time as Darwin, and held that that must suffice for explaining evolution. When Mendel's laws were rediscovered about 1900 and along with it the mode of change in hereditary matter by mutation became known, people felt that it was once and for all mandatory to fix a one-way causality of the evolutionary mechanism.

Not that there was lack of opposition. However, the comfortable mode of thought, our "lazy reason", as Kant would have said, carried the day in the textbooks. Mutations occasionally changed a gene and selection picks out the carrier of the more successful phenotype in each case. A reflux of information was excluded, the assumption of it even forbidden by Weismann's doctrine (since 1902). Herbert Spencer, who coined the phrase "survival of the fittest", roundly spoke against Weismann, but doctrine and prohibition, though surely strange terms in science, have rigidified into a paradigm.

When finally the mechanism of DNA replication and the chemical transfer of messages by messenger RNA and transfer RNA were clarified, the new piece of "truth" was felt as proving the possession of the whole truth. The paradigm of inorganic science, physicists and chemists, that had been so fruitful in this discovery, further cemented the Weismann doctrine into the "central dogma of molecular genetics". What sounds like a mockery of an empirical science might have started as a laboratory joke, but became bitterly serious.

From the dogmatic prohibition even to consider a reflux of information from the gene products became a kind of social truth, and if one did not observe it one was henceforth excluded from the community which allotted means and recognition; debate was silenced, and the heretic became a minority at whose level you must not function precisely because of the ethic of scientific enquiry.

The consequences are thus catastrophic. Pangenesis theory is never mentioned. The whole wealth of phenomena of morphology and trans-specific evolution (the large-scale course of racial history) remains unexplained and is displaced as a kind of anachronistic lumber and finally forgotten. Textbooks agree on a single causal chain, which starts from the DNA molecule and by itself is to explain the world of organisms. The rest is no longer taught, for even the teachers are often unaware of it. Even Darwin is called "a bad Darwinist" by his biographers (Hemleben 1968, pp 100, 129). The whole thing has become absurd. Reductionism from being pragmatic becomes ontological. Biology is on the point of dissolving into chemistry, and so is our view of the world.

Indeed people behave as if only the market, environmental selection, determined the structure of our motorcars for example. The factory tries this and that and buyers judge as to the dimensions of cylinder-head screws, the alloy and tolerances of the piston rings. It would be impermissible to organise the factory, which is to learn only from the market, but not from its own products: no previous experience is to count and indeed the maker should not know his product (Riedl 1975, 1976, 1977).

The modest simplification, by means of which our inborn forms of intuition picture the world for us, are blown up to genuine errors if we extrapolate by rational reflection. Our linear causal thinking blinds us to what is happening in the world round us all the time. The paradigm wins out against reason.

Loss of insight into associative knowledge gain

In the same decades of the early 19th century when the work of Erasmus Darwin (✝ 1802) began to be influential and Lamarck's important book appeared (1809, the year of Charles Darwin's birth), another important idea began to develop; the systemic conditions of our cognitive gain, as we now put it, were recognised.

The materialism or mechanism of the 18th century, of men like Lamettrie, which still impressed Frederick the Great, had proved to be obsolete. The new reductionism arising from the Enlightenment, later to be spread by positivism, was still too young to have impregnated science. In those decades the old-Goethe developed the method of comparative anatomy and the young August Boeckh (1785-1867) that of philological science.

Oddly enough, this happened independently of evolutionary theory, whose rise was mentioned above. Baron Cuvier for example succeeded so well in discrediting Lamarck that Goethe had never heard of the latter, although Goethe was so interested in the controversies at the Jardin des Plantes in Paris (precursor of the biological sections of the University of Paris) that he translated the discussions between Cuvier and Geoffroy Saint-Hilaire into German. Both concepts were thus rooted in the still more general common concept of the turn of the century, with influences from Winckelmann and typical representatives in Schleiermacher and the brothers Humboldt.

Goethe and Boeckh attained the insight into the principle of mutual explanation for the understanding of a complex system, of morphological and literary type respectively. The two men worked quite independently. Boeckh, only four years older than Goethe's son August, never so far as I know refers to the identical method of the master whom of course he was aware of. Incidentally, right to his death in 1867, he based his lectures (published posthumously) on a manuscript finished in 1809, the year of Lamarck's Zoologie Philosophique and Darwin's birth.

Goethe recognised that the type, or basic structural plan of a systematic unit, in his case mammals, cannot issue from any of its representatives. It must always be obtained by abstraction as a theory from all its instances, here from all mammal kinds. Only then can it in turn be used for comparison with each kind of the group and at the same time as an empirical check and correction. Moreover, the type of mammals must at once become one instance with any of the birds, reptiles and so on, from which we can now draw the type of vertebrates.

The same stratified mutual relations are found by Boeckh. The meaning of a so far unknown word will be clear from the sentences in whose sense one finds it, and conversely we know that the sense of a sentence is inferred from the meaning of its words. However, whether a sentence is for example to be understood ironically can not be inferred from the sentence alone. Its deeper sense will emerge only from the instances of sentences with which it is linked in context, just as the sense of the context emerges from the sentences that make it up.

This mutual relation of cognition corresponds to the mutual relation in the self-organisation of things, as the insertions of differentiation between part and whole. For obviously mammals and birds arose between the given whole of vertebrate organisation and the parts or individual species; just as the words and sentences of languages must have developed between the whole of the need for communicating in a group and the parts or possible sounds. This isomorphism between grounds of genesis and grounds of cognition (first in Riedl 1983) was of course not known, but the vision of Goethe and Boeckh was not deflected by the complexity of the mutual relation in cognition. Not so with their successors.

It must once more have been hereditary guidance to the linear form of causal thinking (Riedl 1978/79) which caused a split amongst thinkers towards mid-century. Our endowment to reckon now with material and now with final causes must always end in a conflict with the intuitive form of individual solutions. It is then more the spirit of the times, a socially agreed conviction and appeasement that takes a guiding hand in the cultural scene. Even after 1900 it remained a universalist and holist world picture that in the dilemma of our forms of intuition kept many a thinker in the position of both these outlooks, but things soon changed.

Positivism began to act, fruitfully for the theoretical support of the natural sciences, especially the exact (inorganic) ones. For it was godfather to pragmatic reductionism, which led these sciences to their great successes. This same scientism, materialist in its consequences, was inadequate for biological science, and the humanities and introduced a new schism between the linear causality of a materialist world view and an idealist one. Where one can recognise morphological types, German idealist philosophers were looking for an authority and unfortunately thought they had found one in Goethe. They interpreted types in the sense of Plato as a priori ideas, where Goethe precisely regarded them as an abstraction from empirical enquiry. Moreover, they interpreted the cause, which Goethe called esoteric, as mysterious. They misunderstood, or did not wish to understand, that Goethe was using a polar pair of concepts: exoteric and esoteric, to distinguish outside from inside causes or grounds. Today we speak of conditions being immanent in the environment or in the system, on the system theory of evolution those that are at the basis of homologies or analogies respectively (Riedl 1975). It is these homologies (essential similarities) that constitute the structural plan or type of a system group. The re-interpretation of morphology into a secret doctrine by German idealism was catastrophic for biology, for the theory of structures became idealist, while that of functions became materialist by opposition. The life sciences were split and discussion across the boundary stopped. The gap remains unbridged to date.

Similarly in the humanities as regards Boeckh's insight into the cognitive possibilities of literary types. The man who is regarded as its new founder is Dilthey since in his introduction to the humanities (1st Volume 1883) accords them a method of understanding as against the method of explanation that marks off the natural sciences becoming dominant at that time. However, Dilthey, though still strongly influenced by Schleiermacher, became professor only in 1866 (Boeckh died in 1867). The new act is thus played two generations later, when little was left of Boeckh's reciprocal causes (which were indeed published only in 1877, by Bratuscheck). For Dilthey's assumption is the German philosophy of life or experience, which involves the view that our understanding attains its full scope only in the products of man, because we ourselves are the producers.

This prepares a change from pragmatic interpretation of each special science to a philosophic interpretation. This split will again have serious consequences because over the lowlands of empirical science it allows a celestial vault of speculation that need not fail through experience. Soon hermeneutics is no longer viewed in abbreviation as an auxiliary science but as the "shared critique of the concept of method in modern science". Heidegger, Gadamer, Habermas and Apel hand on this view which observes idealist, existentialist or phenomenological interpretations in positivists and analytical philosophers like Albert, becoming "a substitute for theological plans for salvation" in order to preserve traditional self-designs.

The result is that once again with hopes of mono-causal interpretation, the theory of mutual relation, as contained in the pragmatics of the unhappy concept of the "hermeneutic circle", has all sorts of wrong questions put to it.

For a start: might it be a logical circle? If it were a circular inference, where, next, would we have to begin questions in a circular path? If we knew that, since we seem to have circle upon circle, from which of them, thirdly, would analysis and the original ground have to begin? Since, fourthly, we cannot yet exclude the circle's end (that is, the problem of delimitation), where would analysis end and thus provide certainty?

Those familiar with systemic connections will notice that we here touch on a whole set of systemic properties and pose questions that prevent insight into their connections. If we ask for the developmental laws of self-organising processes which govern the rise of a structure of organisms or communication, we will always have to reckon with reciprocal connections and never with a place of original certainty.

Likewise for the cognitive process of self-governing systems. However, no feeling of certainty for justifying such an expectation has arisen. On the contrary, special disciplines have become unsure of themselves through the development here described. Neither the dialectic of idealist logic nor the formalised logic of analytic philosophy has provided any support in this matter.

The solution is only beginning to appear in outline (Riedl 1984). It rests on the insight that the subsuming schema of regularity, on the basis of symmetrical conditions under which all differentiation arises, requires a mirror-like extension.

This subsuming schema of Hempel-Oppenheim (1948) states that the explanation of regularity subsumes itself into a hierarchical system. Indeed any individual law contains not its explanation but describes a correlation. It will be felt as explained only when with further laws it can be described as a case of a super-ordinated law; for example, the laws of levers and free fall under those of terrestrial mechanics, Galileo's terrestrial and Kepler's celestial mechanics under Newton's theory of gravitation.

The necessity of expecting a double hierarchy can be clarified in terms of our philological example of the starata of words and sentences. The theory of the meaning of a word is developed from all the sentences in which it figures. The theories of meaning of many words at the same time constitute cases of a theory of meaning of signs in the direction of the next layer below. Just so one counterdevelops the theory of meaning of a sentence from the cases of the meaning of the words it contains. The theories of meaning of many sentences at the same time constitute the cases of the theory of contextual meaning in the direction of the next layer above.

In general, for all complex self-organising systems the function, purpose or sense of any layer is to be understood from the selection conditions of the next layer above, from the hyper - system. These are the conditions of formal and final cause discussed earlier. Conversely, the structure, differentiation, operation and impulse of each layer are to be understood from the conditions of availability in the next layer below, from material and efficient causes.

A retrospective glance at complexity

Case studies of this kind can easily be multiplied. We all know of such cases ourselves. They support the general observation that we make the matter of understanding complex self-organising systems too easy for ourselves. Nevertheless our attitude will remain ambivalent; we shall find the more complex solution both correct
and confusing. That is my theme.

When we speak of self-organisation, we expect to have corresponding forms of intuition to picture the matter, which otherwise could not be imagined, as we say;
indeed it would be unimaginable. However, it can be shown that for many phenomena,
that are decisive here, we cannot have inborn forms of intuition. For the self-
organising process that organised the foundations of our reason was not concerned
with the ability to have insight into self-organising processes. It has developed
the preconditions of our reason for the solving of life problems in an as yet
rather simple environment.

Therefore it would not justify reservations, if our power of imagination cannot
picture a putative phenomenon. If we can trust anything, it is only the constant
confirmation of forecasts from our theories, whether supported by intuition or
not. Indeed, the latter case is a challenge of special importance. It offers us
the chance to transcend our own sense equipment and thus to enter into a new phase
of self-organising processes such as evolution.

Let us acknowledge that our senses as theories, as aids to decision and survival,
were organised in an as yet simple domain of problems, and as Ditfurth says, not
for the purpose of epistemology. We further acknowledge that our present complex
civilisation simply happened to us: we stumbled into it, as von Hayek has taught.
What we will not acknowledge is the prospect that because of adaptive deficiencies we are in the process of ruining our world and with it ourselves.

References

Albert, H. (1968):   Trakat über kritische Vernunft, Mohr, Tübingen

Ashby, W. (1952):   Design for a brain, Wiley and Sons, Chichester

Berger, P./Luckmann, T.   Social Construction of Reality: A Treatise in the
(1966):   Sociology of Knowledge, Irvington Publishers Inc.,
                             New York

Bertalanffy, L.v. (1968):   General system theory. Foundation, development,
                             application, Braziller, New York

Boeckh. A. (1966[2]):   Enzyklopädie und Methodenlehre der philologischen
                             Wissenschaften, I. Formale Theorie der philologischen
                             Wissenschaften, Neuausgabe (von 1877), Wiss. Buch-
                             gesellschaft, Darmstadt

Brunswik, E. (1957):   Scope and aspects of the cognitive problem, in
                             J. Bruner et al. (Ed.): Contemporary Approaches to
                             cognition, Harvard University Press, Cambridge

Campbell, D. (1966):   Pattern matching as an essential in distal knowing,
                             Holt, Rinehart and Winston, New York

Churchman, C.W. (1979):   The Systems Approach and its Enemies, Basic, Skills
                           Unit., Cambridge

Darwin, Ch. (1868):   The Variation of Animals and Plants under Domesti-
                           cation, 2 Bde. AMS Press.

Dessauer, F. (1958):   Naturwissenschaftliches Erkennen, Knecht, Frankfurt/M.

Dörner, B. (1975):   Wie Menschen die Welt verbessern wollten und sie dabei
                           zerstörten, in: Bild der Wissenschaft, p. 248-253

Dilthey, W. (1883):   Einleitung in die Geisteswissenschaften (neue Auflage
                           1933, Teubner, Stuttgart)

Ditfurth, H.v. (1976):   Der Geist fiel nicht vom Himmel. Die Evolution unseres
                           Bewusstseins, Hoffmann und Campe, Hamburg

Eigen, M. (1971):   Selforganization of matter and the evolution of biolo-
                           gical macromolecules. Naturwiss. (58): 465-522

Foerster, H.v. (1962):   On Self-Organizing Systems and their Environments, in:
                           Yovits/Cameron (Eds.), Self-Organizing Systems, p. 31 ff.

Goethe, J.v. (1790):   Morphologische Schriften, Böhlau, Weimar

Haken, H. (1981):   Erfolgsgeheimnisse der Natur. Synergetik: Die Lehre vom
                           Zusammenwirken. Deutsche Verlags-Anstalt, Stuttgart

Hassenstein, B. (1979):   Wieviele Körner ergeben einen Haufen? Bemerkungen zu
                           einem uralten und zugleich aktuellen Verständigungs-
                           problem. in: A. Peisl u. A. Mohler (Eds.), Der Mensch
                           und seine Sprache: 219-242. Propyläen, Ullstein,
                           Frankfurt/M.-Berlin-Wien

Hayek, F.v. (1979):   The Counter Revolution of Science Studies on the abus
                           of reasons, Liberty Press, Indianapolis, 2. Ed.

Hemleben, J. (1968):   Charles Darwin; in Selbstzeugnissen und Bilddokumenten,
                           Rowohlt, Hamburg

Hempel, C./Oppenheim, P.   Studies in the Logic of Explanation. Philosophy of
(1948):   Science 15: 135-175

Jantsch, E. (1980):   The Self-Organizing Universe: Scientific and Human
                           Implications of the Emerging Paradigm of Evolution,
                           Pergamon Press, Oxford

Kullmann, W. (1979):   Die Teleologie in der aristotelischen Biologie.
                           Aristoteles als Zoologe, Embryologe und Genetiker.
                           Sitzungsber. Heidelberger Akademie d. Wiss.,
                           Philos.-histor. Klasse, 2. Abhandlung, 1-72

Kullmann, W. (1982):   Wesen und Bedeutung der 'Zweckursache' bei Aristoteles.
                           Berichte zur Wissenschaftsgeschichte, 5, 25-39

Lorenz, K. (1977):   Behind the Mirror, A Search for a Natural History of
                           Human Knowledge, Taylor Roland, Harcourt Brace Jovano-
                           vich Ltd., London

Malik, F. (1979): Die Managementlehre im Lichte der modernen Evolutions-
theorie, in: Die Unternehmung, Nr. 4, Bern 1979,
S. 303 ff.

Maturana, H./Varela, F. Autopoiesis und Cognition, Dordrecht-Bosten
(1980):

Popper, K. (1972): Objective Knowledge, An evolutionary approach,. Oxford,
Clarendon Press

Prigogine, I. (1955): Introduction to thermodynamics of irreversible process,
Thomas, Springfield

Probst, G. (1981): Kybernetische Gesetzeshypothesen als Basis für Gestal-
tungs- und Lenkungsregeln im Management, Haupt, Bern-
Stuttgart

Remane, A. (1971[2]): Die Grundlagen des natürlichen Systems, der vergleichen-
den Anatomie und Phylogenetik. Koeltz, Königstein-
Taunus

Riedl, R. (1978): Order in living organisms, John Wiley, London

Riedl, R. (1984): The Biology of Knowledge, John Wiley, Sons, Chichester

Riedl, R. (1977): A systems-analytical approach to macro-evolutionary
phenomena. The Quarterly Review of Biology (52):
351-370

Riedl, R. (1978/79): Ueber die Biologie des Ursachen-Denkens. Ein evolutio-
nistischer, systemtheoretischer Versuch, in: H.v. Dit-
furth, (Ed.): Mannheimer Forum 78/79: 9-70

Riedl, R. (1980): The Biology of Knowledge, John Wiley and Sons,
(Eng.trans. 1984) Chichester

Riedl, R. (1983): Evolution und Erkenntnis, Piper, München-Zürich

Riedl, R. (1984): Biologie des Erklärens und Verstehens, Parey,
Hamburg-Berlin (in preparation)

Simon, H. (1962): The architecture of complexity. Proc.Am.Philos.Soc.
106 (6): 467-482

Snow, C. (1959): The Two Cultures, Rede Lecture, Cambridge, CUP

Spaemann, R./Löw, R. (1981): Die Frage wozu? Geschichte und Wiederentdeckung
des thelologischen Denkens, Piper, München

Ulrich, H. (1972): Systemorientiertes Management, Management: Intuition
oder Information? Separatdruck Univac 1972

Varela, F. (1979): Principles of Biological Autonomy, North Holland Publ.,
New York

Waddington, C. (1957): The strategy of the genes. Allen and Unwin, London

Wallace, A. (1891): Darwinism, An Exposition of the Theory of Natural
Selection with some of its Applications, AMS Pres.

Watzlawick, P. (1984):   Invented Reality, Norton Publ., New York
(Ed.)

Weismann, A. (1902):   Vorlesungen über Diszendenztheorie, 2 Bde.,
                          Fischer, Jena

Weiss, P. (1971):   Hierarchically organized systems in theory and
(Ed.)   practice, Hafner, New York

Towards a Theory of Social Systems: Self-Organization and Self-Maintenance, Self-Reference and Syn-Reference *

P.M. Hejl

Universität GHS Siegen, FB 3, Postfach 10 12 40, D-5900 Siegen, Fed. Rep. of Germany

The vagueness of the term "science" is a perpetual source of confusion. Its meaning encompasses the traditional empirical, experimental, and formal academic disciplines, e.g. physics or mathematics, as well as most diverse activities which "scientists", who coopt each other, call "scientific". In order to discuss theoretical problems of social theory, the notion of science must first be made explicit. Following a proposal by Humberto R. Maturana (1978a) I define as "scientific" any activity that obeys the scientific method. The scientific method involves roughly four steps. These are:

"1. Observation of a phenomenon that henceforth is taken as a problem to be explained.
2. Proposition of an explanatory hypothesis in the form of a deterministic system that can generate a phenomenon isomorphic with the observed one.
3. Proposition of a computed state or process in the system specified by the hypothesis as a predicted phenomenon to be observed.
4. Observation of the predicted phenomenon." (ibid. 27)

I shall proceed without further discussion of the implications and consequences of this understanding of science (cf. P.M. Hejl 1982b, 47 ff.). As indicated by the title of this contribution, it will deal only with the first two steps of the scientific method, a limitation which will be overstepped sometimes by examples and conclusions.

I

Since the publication of T. Parsons' "The Social System" (1951) social scientists and laymen alike speak of systems whenever they observe the phenomenon that individuals in groups behave in ways different from their behaviour in isolation. As this is most salient in the case of rigid formal structures like bureaucracies or firms, the connection between the notion of "system" and the Durkheimian understanding of the social in general was somewhat lost, although it still exists in the Parsonian tradition. The interdisciplinary "systems movement" with its emphasis on "general laws" of systems behaviour also helped to shape the idea of "the system" as did the much older tradition of treating groups as supra-individual actors. In this view "systems act purposefully", "adapt themselves to goals or to an environment", and, finally, "have problems to survive". Social systems are seen

---

* This is a completely rewritten version of my original contribution to the research meeting. It takes into account some of the results of the critical discussion of what I presented and incorporates new ideas. I want to express my gratitude to the organizers of the research meeting in St. Gall. The particular combination of efficiency and friendliness made the meeting an outstanding event.

as entities of a special kind by all those who consciously use the system's terminology. Consequently there are many efforts to analyse the behaviour of systems, to design and to control them. As the "systems position" is usually "holistic", little or no work is directed at a better understanding of the processes leading to the *formation* of what is labelled "system" and "system behaviour".

During the last ten or fifteen years, furthermore, much thinking in several disciplines has been devoted to gain more clarity about some of the processes we usually describe with the help of the prefix "self-". There is of course "self-organization" (1), perhaps the most famous representative of the scientific branch of this family of expressions whose meaning is still so vague that the traditionalist hesitates to give them the honorary title of "scientific term". But one should not look down upon "self-maintenance", "self-reference" and "self-production". Despite these problems, the ideas these expressions represent are felt to offer such a challenge that it is not only understandable that some of the most innovative scientists have engaged themselves to give them a more precise meaning, but that it is also astonishing that our scientific communities are not more fascinated by the questions behind these expressions.

At the very core of these questions is the problem of how to understand the process of life and hence the functioning of living systems. This question is of major importance both for systematic and for historical reasons. If one looks at the history of science and at social history of which the former is a part, one may speak in an epistemologically somewhat obscuring way of a "hidden research program". Modern science resulted from the growing belief at the end of the Middle Ages that neither revelation nor tradition can provide a firm basis of knowledge. Once this conviction has become sharpened, the logical implication is that man is the sole source of human knowledge and that he can rely only on his own judgement for certainty. This apparently circular situation could be ignored as long as science was more or less identical with the natural sciences. But as soon as we make ourselves the subject matter of science in this understanding, this long disregarded problem becomes virulent, finally splitting the scientific world into natural sciences and humanities (Geisteswissenschaften), all generating a great variety of different understandings of man and science. The hidden research program is the elucidation of the concept of "self" as it is used in psychology as well as in other disciplines. It indicates some kind of autorelation either of an human actor, or of a social group, or of all kinds of "natural" processes.

I want to propose an answer to the following question: Is it possible to unite the systems approach and current work from the research program on the "self-" in a systematic and not merely metaphorical way such that the resulting systems approach may be useful to the social sciences as a whole?

I·I

As the required model of social systems has to generate the phenomena we are interested in through the interaction of its components I shall proceed as follows:

1. The active components of social systems, individuals as living systems,
   will be introduced through a series of definitions; by specifying as precisely

---

(1) The term "self-organization" is in fact the link between the present discussion and that of the early sixties (cf. M.C. Yovits/S. Cameron (eds.) 1960, H. von Foerster/G.W. Zopf (eds.) 1962) which was part of the "cybernetic movement". That this connection is not merely incidental is demonstrated by the rôle Heinz von Foerster, founder and director of the Biological Computer Laboratory, played in the early discussion and, with his proposition of "second-order cybernetics", plays in the discussion today (cf. H. von Foerster 1981).

as possible the corresponding processes, I shall try to overcome the vagueness
of terms like "self-organization", "self-maintenance" and "self-reference".
I will equally show the relation of these terms to the notions of "autopoiesis"
and "autonomy" (as conceived of by H.R. Maturana and F.J. Varela). By doing
this I hope to reach an understanding of "living systems" which is explicit
enough to shed light on similarities and differences between living systems
and social systems.

2. In the second stage, the social domain is generated as a result of the
   interaction of living systems. Then some problems of the theory of evolution
   and their relation to the social domain will be discussed.

3. In the last section a definition of "society" and of "social system" is
   proposed. Then the problem of how to characterize social systems is discussed.
   The main question to be dealt with turns out to be that of the shortcomings
   of the terminological apparatus used for living systems when applied to
   social systems. To remedy the resulting inconveniences a specific term to
   qualify social systems is proposed and discussed.

II.I

As I have shown elsewhere (cf P.M. Hejl 1982a, b) it is necessary to include human
beings in the definition of social systems. This will be substantiated in part 3.
Here it may suffice to point out that even this decision leaves us with "components" of the systems we are interested in, which are far too multi-faceted to
allow for a clear understanding of the consequences of this decision. The clarity
we are interested in can be gained if we provisionally look at "individuals" not
as representatives of the socio-cultural process which produced this notion but as
singluar living systems. I know, of course, that most social scientists are shocked
when reminded that social phenomena are not independent of the biological base that
we are trying to ignore for all too traumatic historical reasons. But this aversion
should not prevent us from scrutinizing biological findings for their relevance to
the social sciences.

As social scientists admit that we (as social beings) are incapable of doing anything that we cannot do as biological systems (with a social mode of life), it
should go undisputed that we have to know how we "function" as biological systems
and how biology and sociality are interlinked. This positively demands a better
understanding of the characteristics of living systems, i.e. of the class of
systems to which we belong. They can be described at the physico-chemical level by
the terms "self-organization", "self-maintenance". and "self-reference". The following definitions or explanations shall state as plainly as possible what is
meant by those terms and in which way they are used here. (1)

Self-organizing we call *processes (or systems)* which, due to certain initial and
limiting conditions arise spontaneously as specific states or as sequences of

---

(1) These definitions were first presented by G. Roth at Bielefeld university in
October 1983 (cf. G. Roth 1984). They are the fruit of many discussions within
a group of friends and colleagues who have been working together to gain a better understanding of living systems and the phenomena they produce. The central
idea uniting this group (U. an der Heiden, mathematician; P.M. Hejl, political
scientist and sociologist; W.K. Köck, linguist; G. Roth, philosopher and
biologist; H. Schwegler, theoretical physicist) is the project of an empirical theory of cognition. Following a suggestion by H.v. Foerster the group
embarked on a closer examination of the theory of autopoietic systems by
Humberto R. Maturana (1970a, b, 1974, 1978a, b (with F.J. Varela) 1979) and

states. Such states or sequences of states can be understood as attractors in the sense of the formal theory of dynamic systems.

An *example* is the highly complex three-dimensional structure of a protein molecule, e.g. of an enzyme, which forms spontaneously as soon as the required components, e.g. aminoacids, are present in the required sequence. Another example is the well known Zhabotinsky reaction.

A self-organizing system is by itself not self-maintaining. This is because its components decompose or are consumed in the process and because there is no possibility to resynthesize or to replace them.

This problem is solved in *self-maintaining systems*. They consist of cyclical concatenations of self-organizing systems such that the first self-organizing system produces exactly the conditions for a second self-organizing system (or process) which in turn produces the starting conditions for a third process ... until one of the self-organizing systems produces the initial conditions for the first system in the cycle. (To be quite precise: for a system belonging to the same class as the original system which "started" the cycle.) Self-maintaining systems are systems in which self-organizing systems "produce" each other in an operationally closed way. To state it differently: self-maintaining systems are systems whose components maintain each other, and by maintaining each other uphold the whole cycle.

Examples are to be found in the metabolism of cells as well as in the interactions between the components of an organism.

Self-referential systems are systems which organize the *states* of their components in an operationally closed way. This leads to the consequence that

- self-maintaining systems are necessarily also self-referential, but
- not all self-referential systems are self-maintaining.

If one looks at an organism, for *example*, it is clear that the circular production of its components includes the circular production of the states of the components. At the same time, we have to take into account that the brain as a self-referential system (cf. F. Varela in this volume) does not maintain itself. The brain as a self-referential system is characterized by neuronal activity which leads to neuronal activity. It is not this neuronal activity which directly maintains the brain. Its maintenance is assured by the organs of the organism to which it belongs. (1)

---

his former student and present colleague Francisco J. Varela (1979, 1981, (with H.R. Maturana/R. Uribe 1974) as a most promising approach to tackle the problem of "self-" in biology as well as in cognition theory. The work of the group is documented in numerous articles and several other publications, e.g. P.M. Hejl/W.K. Köck/G. Roth (eds.) 1978, F. Benseler/P.M. Hejl/ W.K. Köck (eds.) 1980, G. Roth/H. Schwegler (eds.) 1981, P.M. Hejl 1982a. From this background it should be clear that I am indebted to the colleagues named above whom I thank for years of continuing discussion and that therefore the ideas presented in this article are only partially "mine", although I am of course alone responsible for the shortcomings of my presentation.

(1) This has the consequence that we have to consider the brain
    a) as an organ - when we ask for the rôle it plays in the self-maintenance of the organism (and hence itself) - and
    b) as a self-referential system - when we ask how it functions.
    The "connection" between these two ways of looking at the brain seems to be its topology. As a result of its evolution the brain represents in its topology the sensorimotor surfaces of the organism such that its activities, although self-referential, are necessarily linked to the organism (cf. G. Roth 1980, 1984).

These definitions allow us to understand living systems (organisms) as self-main-taining concatenations of self-organizing processes. By distinguishing between the production of components and their states, living systems become self-maintaining and self-referential. They can provide the material basis of systems like the brain, which (with respect to cognitive functions) are self-referential but not self-maintaining. This explanation of living systems is understood as an attempt to give a more explicit meaning to the notion of an *autopoietic system*. Particu-larly the distinction between self-maintenance and self-reference allows us to separate phencmena which otherwise are too easily included in the same category (cf. N. Luhmann 1982). It will be shown that this is of great importance to the understanding of social systems.

Whereas the terms discussed above can be explained by describing the corresponding processes or systems, this is not the case with the term "*autonomy*". As used by F.J. Varela (1979, 55), it seems that "autonomy" is primarily understood as a pro-perty of operationally closed systems. This engenders the possible confusion of autopoiesis (as defined above) with autonomy. To avoid this danger I propose to use "autonomy" as a term which designates the particular experience an observer may wish to express when describing a living system with respect to what he takes to be its environment.

This experience is in fact, as has frequently been shown by H.R. Maturana and F.J. Varela, due to the operational closure of living systems. Operationally speaking, self-maintaining and self-referential systems do not have inputs. Their environ-ment constitutes a source of perturbations or modulations for the processes which constitute the system. The result of a given perturbation or modulation depends on the state of the system (or subsystem), on its "structure", to use the terminology of the theory of autopoietic systems. This has the effect that the reactions of the system to what an observer takes to be analogous inputs may be of great varie-ty. *This* input-independence of living systems I propose to call "autonomy". If autonomy is understood in this way, it positively designates the relation between the system's state, hence its environment, and the observer's environment. Autonomy therefore does not imply autonomy from the environment in a solipsistic understan-ding or in the sense of self-sufficiency, but affirms the epistemological limita-tion of all cognition, namely that all knowledge is observer-dependent.

This becomes clearer if we ask what happens if the input approach is used. To speak of "input" implies the "breaking up" of the operational closure of living systems. As a result, the latter becomes an operationally open system which is part of a more encompassing system, e.g. including parts of the environment. In such a model, the behaviour of the system is determined by the input from the environment. Unfor-tunately, in order to be successful such a view presupposes identical knowledge about the universe on both sides, on the side of the observer and on the living system. Both need to "know" the same environment (even if the environment of the observer may be larger). If this is not the case, the autonomy of the system stands against the autonomy of the observer. As the rôle of the observer is normally not taken into account explicitly in our scientific tradition, the observed system is often blamed for the resulting incongruencies. As an illustration, notions of *optimality,* when applied to the relations of living systems (and social systems) to their environment, are typical of this situation. Speaking of "optimal adapta-tion" for example has every appearance of a statement not about the relations bet-ween a living system and its environment (see below), but about the relation bet-ween a living system and what *an observer takes to be* its environment. Such a statement does not increase knowledge about the observed system but about the ob-server.

Every time an observer tries to predict or to control the behaviour of a living system on the basis of the only available knowledge he has at his disposal (earlier observed behaviour and his conclusions about the system's environment), autonomy appears as the degree of the system's "freedom" due to the differences between the system's state (and hence environment) and the observer's inferences about them, and thus between the system's behaviour and his expectations. As a slight shortcoming of the proposed explanation it should be noted that despite the fact that much stress has been laid on the observer and his predictions, the phenomenon of autonomy is produced by the operationally closed organization of living systems. Considering systems with highly complex self-referential subsystems, like ourselves, one should add as a further source of autonomy the apparent capability of these systems to generate *new* internal relations. It seems that this capability leads to the generation of the phenomenon of innovation and creativity which surely is an important feature of autonomy.

At this point a short résumé is necessary. So far living systems (organisms) have been characterized as autopoietic systems through an explanation of "self-organization", "self-maintenance", and "self-reference". The term "autonomy" was proposed to denote the freedom of a living system with respect to an observer. It was further pointed out that these explanations (except that of autonomy) describe the living system as a physico-chemical system.

As shown by the different examples, no specific difference was noted between very elementary systems at the molecular level and organisms with a brain up to systems of our own complexity. This should not constitute a major problem as long as differences due to different degrees of complexity are not overlooked. Not surprisingly, the possible "performance" of a system depends on its complexity, namely its internal differentiation and specialization. This is not tantamount to saying that what has been said about living systems up till now suffices to understand *human social systems*. Their comprehension requires a more detailed discussion of the function of the brain, on the one hand, and of the social, on the other.

II.2

The human brain and our social way of life seem to be more deeply connected than is commonly assumed. This will be shown by an examination of some aspects of the theory of evolution (1) in the first half of this section. In the second half I will try to show how the phenomena which in the first half are inferred from the theory of evolution may be generated.

As is well known, evolution is explained by the theory of evolution as resulting from two interlinked processes: genetic variation and selection. In this context, selection is seen as an optimizing process whereby fitness increases and leads to better adaptation.

This "adaptationist" view encounters problems which shall be discussed successively:

1. Not only epistemology but also biological perception research demonstrates that system-independent perception of what is currently called "objective reality", in our case "environment", is impossible.

2. There are living systems which have not undergone morphological changes for more than 50 million years during which the world has changed tremendously (cf. D.B. Wake/G. Roth/M.H. Wake 1983), a striking example of the phenomenon of stasis.

---

(1) I am indebted to G. Roth for discussions on the matter of evolution which helped to clarify my ideas about the biological/systems-theoretical explanation of society (cf. G. Roth 1982, D.B. Wake/G. Roth/M.H. Wake 1983).

3. There are living systems which changed although their environment (in the objectivist sense) did not undergo alterations.

D.B. Wake/G. Roth/M.H. Wake (op.cit.) conclude on the basis of their own work on salamanders of the family Plethodontidae and of other empirical findings that the theory of evolution must be enlarged by the theory of autopoietic systems if it is to gain explanatory depth. What seems to happen in evolution is primarily a change (modification of old/formation of new organs including brain growth,etc.) in the living systems themselves. Each evolutionary step "defines" (see below), just like every state of the system, a possible range of new interactions, hence a new niche. If this niche can actually be "realized", i.e. if the living system is able to maintain itself (its autopoietic organization) through interactions according to its new possibilities, the system will survive, and a step in evolution will have been taken (cf. above "autonomy"). This explains why objective knowledge is not necessary for a living system, why stasis is possible, and why environment-independent changes occur. Moreover, there are changes which are not only a kind of "shift" but which are additional in that they broaden the sensorimotor possibilities of the living system. This allows for the occupation of a larger niche or for survival if the environment changes in a way that does not permit to secure the maintenance of the autopoiesis in the previous manner, but in a manner already defined by an earlier evolutionary modification.

In this perspective, *evolution* becomes the successful generation of new niches (realities, environments) through internal changes. The environment does not play a positive but a negative selecting rôle by limiting evolutionary changes to those which allow the system to persist. Thus "successful" in the above statement means only that the system survives.

There is another group of problems which the theory of evolution hardly explains but which, at a general level, pertains to its domain: the evolution of human culture and language.

1. Although we do not know when exactly hominids separated from pongids and when human culture and language became specific, we have to assume that our ancestors did well for enormous stretches of time without them.

2. As a changing environment does not provoke specific genetic changes, there is no causal relationship between environmental changes and the evolution of these capacities. It follows that they must have been brought about by the systems themselves.

The conclusion that our ancestors developed the capacities we are interested in (not specific relations like a given culture and language!) as a result of evolution in the sense described above, seems to be backed from two sides. Fossil records show indeed a continuous morphological change with an important growth of the brain in size, especially of the frontal lobe. This is of course only a rough indicator but we have to assume that the human capacities are - to say the least - not independent of the human brain and its capacities.

If we understand the sensorimotor apparatus of a living system as defining its possible cognitive domain, and hence the possible environment in which it is able to maintain itself, the human living system defines in fact an extremely wide range of possible niches. This has led biologists and anthropologists to the well-known statement that man specialized on adaptability. The specific organ for adaptability, however, is the brain.

If we look at the phylogeny of the vertebrate brain, this specification becomes obvious. There is a quite distinct evolution from primitive nerve systems, which merely connect motor to sensory surfaces, to complex central nerve systems whose capacities lie not so much in a better sensory "input" but in the increase of the internal capacities to interconnect and compare the different states of the system modulated at the periphery (hence from "outside") of from within. As this point is very important to my argument I want to cite just one example (for which I thank G. Roth): The optic nerve of the frog, which is the visual entry to the brain, consists of approx. 500 000 fibres compared to man's one million. But whereas the frog's brain has just a few million nerve cells, man has approx. a thousand billion. This demonstrates what specialization on adaptability means: growth and complexification of internal capacities to generate new realities.

Thinking about the consequences of substantial increases in brain capacity, one question becomes of great importance. As one has to assume that like every evolutionary step, the growth of the brain takes place due to still widely unknown interior factors and processes, and is, of course not due to the reactions to changes in the environment (in the objectivistic sense), the system must live in a stable self-generated environment which it can handle without the new possibilities. The question now is: What happens if a living system in a relatively invariant environment undergoes an important rise in his capacity to generate relations between the states of the subsystems, which are partly modulated from "outside" and partly from those regions which,e.g.,generate the phenomena of memory of interferences of any kind? It seems obvious that there must be a "point" in the evolution of brain capacity, where its growth causes the generation of the actual environment (the only reality of the living system) to become increasingly contingent (cf. P.M. Hejl 1982a, 331 ff.). There are two consequences to be considered:

1. If we stipulate that there is a close connection between the niche (or the reality as it is defined by the system) and the behaviour of the system (which has to be assumed the more strongly the less the capacity of adaptation is developed), it becomes clear that the appearance of contingency loosens this connection: the selection of adequate behaviour must become difficult and riskful. A system that is not sure which reality to construct from the few signals it gets from what is perceived as the "outside" is of course not able to define what "adequate" means with respect to the behaviour it has to choose. The result is that the growth of the brain as such constitutes in fact a source of *danger* for the living system concerned.

2. If, on the other hand, a less specialized system gains in capacity to generate more realities under proper conditions, it is far better prepared to secure its maintenance (its autopoieses) through what an observer may call "occupation of new niches" than a system without this capacity. Therefore the growth of the brain in size constitutes a potential *advantage* for such a system. The resulting question runs: How is it possible to avoid the danger without losing the advantage? The answer was, I believe, the "invention of society". This "invention", undoubtedly a long evolutionary process, not only allowed for the channelling of the potential dangerous effects of our growing self-referential brains through the "invention" of myths, religions, art, and finally science, but it also turned the dangers into factors of coherence. By combining (and partly supplanting) individual definitions of realities with social ones, it became possible to secure biological maintenance and, at the same time, to provide domains where the self-referential capacities of our cognitive systems could display their innovative potential.

To resume these considerations, one can state that man is social for biological reasons *and* biological as he is because he is social. Society as such is thus biologically necessary. This does not mean in any sense that specific social regulations, norms, institutions, or socially defined realities are biologically necessary. This excludes any biologistic approach to explain particular social phenomena. That reductionism and biologism are excluded by the proposed model will become plain through the attempt to generate the social from the interactions of living systems and from the discussion of social systems.

When speaking of evolution, great emphasis was laid on reality or environment as the results of a process of construction or definition. In this context it was said that a system "defines" its reality itself. This "definition" is, of course, not a verbal one but due to the functioning of this class of systems. A system without input is a system whose "representation" of the outside depends on his own state which is modulated by the influences its external surfaces undergo as a result of what happens outside. The modulated state of the system *is* the biological representation of his experience(s), his perception. As pointed out already, the fact that we interpret some modulations of our cognitive system as "hearing" and others as "seeing" is the result of the topological structure of our brain which interprets excitations of the visual or auditory parts of itself as resulting from corresponding activities in the related sensory surfaces (the eye or the ear). Therefore it is quite understandable that excitations of those areas produce the illusions of hearing or seeing although for the eye or the ear nothing happens. As in the "normal" situation such excitations happen mostly as the result of external events, it is easy to understand that the self-referential character of the brain is not only no obstacle to its effective functioning as an organ but even the only way to relate the necessities of the system to what an observer may describe as the manifold possibilities of the world. The self-referentiality of the brain explains the "selective behaviour" of the organism. (From what has been said it should be obvious that "selective behaviour" does not refer to the reality of the system but to the reality the observer assigns to the system.) The state-dependence of any perception of a living system is the reason why these perceptions must be regarded as "non-objective" in the traditional sense and therefore effective for the system's self-maintenance. It is this relation between the living system and the reality it generates or constructs (in the sense of E. von Glasersfeld) that has been called "definition", hence "definition of reality".

Just as the sensory apparatus of a living system defines its cognitive potential, its motor apparatus defines its possibilities to handle its environment. Both are linked insofar as a system can perceive only those effects on its actions which modulate its own sensory system. Therefore one has to conclude that the motor-sensory apparatus of a living system defines its possible range of experiences and actions. But living systems do not live in a *possible* but, of course, in an *actual* reality. At this point the difference between the production of components and that of states of components comes in again. It is the state of the system at a given point in time which determines the actual reality of the system. Every interaction of the system modulates its state (or leads to its regeneration, which gives rise to the phenomenon of "objectivity", cf. P.M. Hejl 1980, 150 f) and thus makes *living systems historical systems and their realities the result of an historical process.*

As no living system can control the result of its external interactions completely, which is mostly due to the presence of other living systems, it continuously undergoes a process of ontogenetic change, which can be understood in the sense of what is called "socialization" in sociology. This process of interactions→ modulation of state→definition of a new reality→modified interactions→further modulation of state etc. generates, on the one hand, a reality of rather stable "objects", and on the other, the perception of other centres of activity of a complexity of behaviour comparable to that of the perceiving system: other living systems of the same complexity. In the latter situation it is no longer possible for the observing system to modify its own states unilaterally such that it may achieve reliable predictions of the other systems. Instead, it becomes unavoidable to join in *a process of mutual interactions and hence modulation which results in a partial parallelization of the interacting systems.* As far as this parallelization is brought about, a *social domain* is generated.

My claim is that this definition of social domains covers all social phenomena inasmuch as there will be no phenomenon which is currently called "social" that cannot be related to this definition.

As a consequence of this definition of the social domain and of what has been said about the definition of reality by living systems, social behaviour and communication are to be understood in the frame of the proposed model as follows.

Social behaviour is *every* behaviour that is generated on the basis of a socially produced definition of reality, or that leads to its formation or modification. This definition of social behaviour is broader than the famous definition by Max Weber (cf. 1976,8),who defines social action as an action "which through the intention (Sinn) of the actor or the actors is related to the behaviour of others and whose course is oriented at their behaviour." M. Weber's definition seems to exclude the behaviour and actions produced by an isolated individual, which are, however, explainable only by resorting to a socially constructed reality. According to the proposed definition of social behaviour, the conduct of Robinson Crusoe is perfectly social, even before the arrival of his companion Friday, because it is clearly based on the technical knowledge and on the moral standards of the English society of his time, that is on a socially developed definition of reality and the ways of handling it. The advantage of such a more encompassing definition lies not only in the differentiation between the formation of, the conformity with, and the modification of, social domains, but as well in the possibility to account for lasting effects on the behavioural and cognitive level of sociality. It allows to include in the domain of social analysis cases like that of Robinson Crusoe, which are by no means limited to the literary field if one thinks of what happens, for instance, when a manager changes the firm he works for, bringing "with" him all the experience gained in other firms, i.e. other definitions of reality; or also the domains of science and technology which, of course, always involve very specific definitions of reality.

The importance of social domains lies in that they allow for coordinated behaviour and for communication. If a living system acts in the way described by a social domain, his acts are interpreted by the other living systems, with whom this social domain was constituted, in the way they were intended. This is necessarily so because the cognitive system functions as a self-referential system. As every perception is interpreted according to the state of the system, it follows that, if several systems generate parallelized states, they will, due to their states, identify every event which belongs to the domain they constituted together (a consensual domain), in an equally parallelized manner. This holds for actions of all kinds. Every action actualizes experiences in the system (or modulates it in such a way that the result can be described as such) which are the result of earlier interactions. When they are due to social interactions, one can say that every action "designates" past social experiences, which, when actualized, become a present social reality. When such designating actions are replaced by socially worked-out symbols, verbal or non-verbal, a language is produced. Its domain of reference is not a reality in an absolute sense but a socially constructed or defined reality: that is, a number of parallelized states in those actors who constitute the specific social domain. For this reason, on the level of communication, it is not possible to be sure that communicative interactions have been successful (cf. on communication W.K. Köck 1980, 1981). The proof of successful communication is not the communicative assertion of the partner, but a behaviour which allows the speaker to conclude that his partner interpreted his communicative effort the way it was intended. If we look at the proposed model of a social domain, it becomes understandable that individuals (still taken as singular actors) behave differently inside and outside a social domain. As long as their actions and communications are situated by themselves in the social domain, they have to make sure not to leave it, i.e. to use actions and symbols as defined by the social domain. Outside a specific social domain, or when discussing topics of this social domain in the context of another social domain, they will behave differently. As in principle, and independent of the fact that in our internally differentiated (but not disconnected) societies a great number of social domains is based on the same definition of reality, every social domain is meant to define its own reality, this difference of behaviour is a normal result of living in our kind of societies. In this sense N. Luhmann (1971) is quite right in pointing out that,e.g.,in the political and in the administrative domain politicians and administrators do not and cannot use the same criteria.

II.3

After the introduction of the notion of living systems and how they generate the phenomenon of the social, the last section of this part shall be devoted to explore the consequences for a theory of social systems.

I define as a *social system* a group of living systems which are characterized by a parallelization of one or several of their cognitive states and which interact with respect to these cognitive states.

Unlike the definition of social domains, the definition of social systems is rather restrictive. The restriction stems from the required interactions. The number of people interested in football (soccer) amounts to several millions in most countries of Western Europe. They more or less know the rules of football playing and how games are organized. According to the definitions given they constitute a social domain but not a social system. A football team, on the other hand, belongs to (or participates in) a social domain and constitutes a social system. In order to speak of a social system it is necessary that all conditions enumerated in the definition be fulfilled. Therefore,e.g. all the football players of a country (or of the countries in which there are football players) do not constitute a social system, called football team, because they do not actually play together. But they may temporarily form social systems of another kind when they play matches. They may even constitute a permanent system, equally of a different type, e.g. through their participation in the international business of buying and selling football players. The requirement that the elements must interact does not mean that they have to interact directly, as is demonstrated by the last example. They may interact indirectly through elements placed in between them. But interaction is just one part of the definition. The other part consists of the necessary parallelization of the cognitive states of the components. If,e.g.,a group of people who do not participate in the social domain of football start kicking a ball around, work out their own rules, and so on, nobody would call the resulting game "football", even if the participants in this game would label it so. As every social domain corresponds with a particular reality that has its ontological base in the individuals who constitute it, it is not through the use of a certain kind of vocabulary that one enters such a domain. In accord with what has been said about communication, it is of greater importance to master the required behaviour. The inventors of the new ball game would soon be rejected by those who belong to the consensual domain of football because the inventors' use of the word refers to a set of actions which is not acceptable with respect to what has been worked out under the label "football" in the greater and previously existing social domain "football".

A problem of particular difficulty is that of the *boundary of social system*. Although this question is more complicated to answer for social systems, it is by no means special to them. In biology,e.g.,there is no difficulty to reach a consensus about the boundary of a living system, an organism for example. Its boundary is the skin which covers its exterior. Yet if one tries to define it in a physical terminology, it turns out to be far less unequivocal, because substances and processes which can be found in the system can be found outside as well. In the end one has to resort to a definition which is based on abruptly changing gradients of

intensities, i.e. to the observer-dependent domain of our possibilities of distinguishing differences in measurement. This observer-dependence becomes even more
obvious when we turn to ecosystems. Where is the boundary of an ecosystem like a
lake? Is it the border of the water surface? Do we have to include a strech of the
surrounding land, and if so how much? If we use gradients, e.g. temperature, there
may be cases where this allows only for very poor distinctions. The best way to
overcome these difficulties is to abandon the search for "natural" boundaries and
to stay within the limits of the scientific method.

The result is that we have to define the system in such a way that all the components which participate in the generation of the phenomenon we wish to explain are
included. The pivotal point of the definition of the boundaries of a system is
therefore the explanation of the problem an observer (or a community of observers)
has chosen. (This does not discard the search for a "best" or "optimal" definition
of the boundary of a particular class of systems as long as the resulting definition is not misunderstood as "natural".)

At this point it seems useful to remember the relation between "component" (and
the very misleading "element") and "system". As it is not possible in the context
of this contribution even to outline the discussion concerning "wholes and parts"
(1), I will just give a definition and some short comments. The *component of a
system* is defined as an entity which is described by the properties the system
builder or constructor attributes to it when constructing the system. This definition may help to recall that "system" means nothing else than "a set of entities
and the relations between them". The relations between components are, of course,
not just *any* relations that may exist or be thought of. The relations which make
an entity the component of a system are only those interactional relations between
the entities through which the system is constituted. (Hence I exclude mere classification systems.) In order for these relations to be established, the entities
must be chosen or constructed in such a manner that exactly this can happen. One
consequence is that it is not possible to analyze a system into parts and to describe all the properties one may find when scrutinizing the result. What one does
in this case is not to examine the components of the system but entities of another kind, of course, with different properties (2). If one looks at a football
player, to go back to this example once more, e.g. as a component of the team (the
system) to which he belongs, we would end up in great trouble if we did not distinguish between his being a component of the football team and his belonging to a
family, as a father for instance. By using the proposed definition the mysterious
relation between wholes and parts can be solved for the domain of living and of
social systems. As "system" means nothing "more" or "else" than "set of entities
and the relations between them", hence "the set of interacting components", a
system cannot display properties which are not due to the interactions of its components; both expressions are interchangeable. The advantage of the term "system"
lies exclusively in the economy it allows because of its shortness. But just as
any mystical "holistic properties" are refuted, no reductionism is possible. We
will meet this problem again later on.

---

(1) For a discussion in a larger biological context cf. G. Roth 1981. For the
    connection of this topic with reductionism in the social sciences see P.M.
    Hejl 1982b, 53 ff. The actual state of this topic is represented in "Parts
    and Wholes" 1983.
(2) There seems to be a frequent and confusing interchange of the words "element"
    and "component". As "element" is often understood in the sense of "smallest
    unit", it is often used to designate components. But unfortunately an "element"
    is of course an entity which can be analyzed as an independent entity. Toget-
    her with the influence of Cartesian thinking, this leads to the misunderstan-
    ding of a component being analyzable independently of the system which it
    constitutes with other components.

For the question of how to define the boundaries of social systems one has to conclude that the boundaries are constituted through the interactions of the components. As in our internally differentiated societies every individual participates in the constitution of a great number of social systems, it is only logical to *understand the individual sociologically* as the result of this "overlapping" of several social systems in the same individual. An individual is therefore at the same time a component of several social systems which are interconnected through such multi-component individuals. This permits us to define *society* as a network of interlinked social systems with the individuals as "nodes". Obviously any observer or observer community is also part of this network. This has the consequence that for any analysis of social systems and hence their boundaries, it is not enough to define these as an external observer. If we want to know where the boundaries of what we take hypothetically as a social system are, we have to observe as well as ask the individuals who constitute it what their interactions are and how they "see" (define) the reality of the system (or the systems) they constitute. The question of boundaries thus turns out to be mainly one of empirical research in the proposed frame of social systems.

What kind of system is a social system? Is it self-organizing? Is it self-maintaining or self-referential? To answer this question a comparison between the explanation of these terms and of what happens in a social system seems to be the best way to come to a conclusion.

To classify a *social system as self-organizing* without further specification would establish an equivalence between the processes in physico-chemical systems and social systems, which does not correspond to what we observe in both cases. The main difference on the operational level (if we neglect the great differences in complexity between the components for a while) seems to be the spontaneity of self-organizing processes in the physico-chemical domain which is without equivalence in the formation of social systems. The constitution of a social system is by no means a rapid and spontaneous process. It necessitates a rather long series of interactions by which the interacting individuals constitute the system and transform themselves into components of that very system. Moreover, in the social domain (taken in the general sense) there are not those quantities of parallelized individuals which would allow for a comparison with substances that under proper conditions organize themselves in such a way as to form highly standardized systems of a particular class. The social system generated through the interactions of a group of individuals is highly specific due to the historical character of the evolution of the cognitive domains of the individuals concerned. So there is not only no spontaneous generation of systems in the sense of physico-chemical systems but the results are also necessarily of greater variety. This point can even be pushed as far as to state that in case of only minor differences between social systems, however generated, this would be a point of special interest and a phenomenon to explain.

To classify *social systems as self-maintaining* seems to be a certain temptation. But such a classification would confuse the differences and relations between an organism and its brain as a self-referential but not self-maintaining system and establish a parallel between organisms and social systems. First, the discussed advantage of the human specialization on adaptability would be eliminated from such a model of social systems. Conceptually, social systems would become less complex than a human organism! Secondly, the old biologism (in distinction from the new biologism, i.e. sociobiology, following E.O. Wilson 1975) in the tradition of H. Spencer would be reasserted, i.e. society would be understood as an organism. Not for ethical reasons alone - the connections of this line of thinking with racism and fascism are well-known - but also on a systematic basis this classification is not acceptable. One of the most important features of social systems is that their components, individuals as far as they participate in the system, [1] *constitute at the same time* a more or less important number of *different social*

systems, and [2] have, at least in principle, the *possibility to withdraw* from a
social systems. Now, components of a self-maintaining system are components of
only one system and do certainly not participate in several systems at the same
time.

At this point an incongruity in the literature should be mentioned. H.R. Maturana
and F.J. Varela (cf. e.g. 1975, 53 ff.) discuss autopoietic systems or higher order. Although both (cf. as examples H.R. Maturana 1980 and F.J. Varela 1979, 53 ff.)
underline that societies or social systems are the result of the interactions of
autopoietic systems but not autopoietic systems themselves, their discussions of
autopoietic systems of higher order gave rise to exactly this interpretation. (1)
This misinterpretation seems to be partly supported by the different views of the
authors as regards higher-order autopoietic systems as well as the existence of
autopoietic systems outside the biological domain.

Of course nobody knows if there are autopoietic systems of a nonbiological nature.
The question can only be answered by the demonstration that such a kind of systems
exists. Such a demonstration has not been given up till now. Living systems all
live exclusively in the physical domain and operate accordingly. The fact that
we are still (and perhaps for ever will be) unable to establish unequivocal relations between neuronal processes and specific events in our phenomenal world does
not allow us to speculate about "other" domains in which there might be other
types of autopoietic systems. We know that without physical living systems and
their cognitive subsystems there are no such domains as the linguistic or the political or the economic domains. Of course we may saddle Pegasus and ride to unknown
worlds, but, taking a somewhat down-to-earth point of view, are we doing more than
using the possibilities our self-referential brain offers?

F.J. Varela rightly points out the difference between autopoietic systems and the
phenomenon of autonomy (1979, 55). This difference can equally be found in the set
of definitions proposed earlier. Autonomy may be the result of the actions of autopoietic systems, be they observed ones or observers, or it may stem from,e.g.,so-
cial systems, i.e. from systems constituted by autopoietic systems and yet different from them. But, to use again a biological example, are cells and multicellular organisms not both autopoietic, hence self-maintaining systems? If the question
is put this way, the answer is clearly positive. But what about a cell as the component of a multicellular organism? As a component a cell is not auto- but allopoietic. Here one has to be very explicit about the aggregational level used. If
we compare a single cell with a single multicellular organism, they are both (in
principle) *self-maintaining as distinguishable unities,* independent of their components. But a component, following the given definition, is described exclusively
by an enumeration of those properties which permit the particular entity to take
part in the constitution of the system and hence become a component. In this sense,
a monocellular system and a multicellular system are systems "in their own right",
independent of how their components are defined as long as this definition explains
the behaviour of the system.

But is the description of a component of an autopoietic system as autopoietic itself compatible with the generation of self-maintenance at the system level? I
think that this is, in fact, the case though only within very narrow limits. If
an autopoietic system constitutes what is called an autopoietic system of a higher order, it has to behave in a way indistinguishable from allopoietic systems.

---

(1) Cf. St. Beer 1975 and N. Luhmann 1981, 33 ff. who uses "self-referential
     systems" as a term to designate the "political system" which is defined as
     "a system ... which produces and reproduces itself the elements of which
     it *consists*". In a footnote, a parallel is drawn between this definition
     and the definition of life by H.R. Maturana and F.J. Varela.

If it ceases to function this way, the concatenation of the processes constituting the system will be interrupted sooner or later. The result is that the description of a component of an autopoietic system as autopoietic in itself may be useful to account for certain malfunctions of the system. They will be thought of as "pathological" with respect to the functioning of the system. This is a consequence of the chosen level of analysis, that of the system. By this decision all properties a component might display when taken not as a component but as an independent unity are subjected to the system; hence the normative importance of the system level. The consequences of understanding society of social systems as self-maintaining in the sense of autopoietic systems of higher order would be a normative theory of social organisms with,e.g.,human rights as a mere function of "social needs". But who is to define them? The social and political consequences need not be explained any further.

This problematic relationship may be elucidated further by answering the question: What is the "self" of a self-maintaining system and what is the "self" of a social system? The "self" of a self-maintaining system is the operationally closed concatenation of self-organizing processes (independent of whether an intermediate level of "organs" is introduced or not). If this concatenation is destroyed, the self-organizing processes will come to an end and the cells will die. If a social system is destroyed, the self-maintaining and self-referential component individuals will live on. According to the kind of system some of them will even be quite happy about the system's destruction, while others may suffer for that very reason. Unlike biological systems, social systems do *not produce* their components (1). They are recruited to, and integrated into, the system by the activities of the component individuals or due to their own initiative. Therefore the conclusion has to be drawn that only living systems can be called autopoietic or self-maintaining. Social systems are of another type.

If social systems are not self-maintaining, perhaps the *classification as self-referential* is more appropriate? As a matter of fact, the classification of social systems as self-referential seems to correspond better than the previously discussed alternatives to the characteristic features of this type of system. In particular, the apparent analogy with organisms seems to be avoided. Moreover, such a classification would underline that the system changes if the states of the components change. This allows us to take into account the relation between the states of components (and of the system as it is defined) and the definition of reality.

Yet a social system is neither an organism nor a brain. Individuals are not neurons which constitute self-referential super brains-like social systems. Unlike individuals who participate in the formation and the working of a social system, neurons "have no other reality" than the one generated within the brain (even if

---

(1) The social system "family" does not produce its components. They are "produced" in a way which, at its core, is not subject to a socially defined reality although the related behaviour and the explanations given are to a great extent social, of course. As far as procreation is concerned, the family is a biological system, as far as the organization of procreation, of child-rearing, and the division of labor in the family are concerned, it is a social system. In the same sense, a firm through which the individuals make their living does not "produce" them. To understand a firm as a self-maintaining system would have the consequence to make it responsible for,e.g.,the unhealthy preparation of food by the employees or for the influenza they catch. Even the obvious danger of all kinds of pollution is no counter argument. Pollution does not produce components but threatens to kill them. It is a political and hence social question what kind of nature we want, how we can achieve it, and what "prize" we are ready to pay.

this reality includes alternatives). Moreover, no neuron has the slightest possibility, not even in principle, to leave the system of which it is a component. Our phenomenological world has to be considered as the result of the whole system, not of any singular neuron or group of neurons. Finally, there is another argument which definitely forbids a classification as self-referential in the way this term has been defined so far. Self-referential systems are defined as systems which organize the states of their components in an operationally closed way, and as the organismic metaphor has been discarded, social systems would have to be understood as self-referential but not as self-maintaining. It has been pointed out, when discussing the relation between organism and brain, that the maintenance of the brain is assured by the organism to which it belongs, whereas the functional integration of the self-referential system "brain" in the organism is secured through its topology. Now, if we would accept the brain as a metaphor for the understanding of social systems, we would have to answer two questions: What is the self-maintaining system which assures the production of the components of social systems?, and: What is the topology of social systems which ensures that they display brain-like functions for this system?

To sum up the discussion of how to characterize a social system, one has to conclude that the *result* of the effort to give a precise meaning to the terminology which is "in the air", is *not to use this terminology for the specification of social systems*.

A characterization of social systems has to take into account:

- Social systems are constituted by living systems that are free to participate or not to participate in their formation. If they do participate, however, they do not lose their character as individuals.

- Human living systems always participate in a varying number of social systems at the same time.

- Unlike self-maintaining systems, social systems do not physically produce their components.

- In contradistinction to self-referential systems, social systems do not organize *all* the states of their components and hence do not determine a system-related reality as the *only* reality which is accessible to the component living systems.

- In contradistinction to the components of biological systems, all components of social systems have direct access to the environment of the whole system.

To avoid any confusion between social and other systems and to underline their specificity, I suggest to term *social systems "syn-referential" systems*. (1) The *definition* of this type of system I propose as follows: Syn-referential systems are constituted by components, i.e. living systems, that interact with respect to a social domain. Thus the components of a syn-referential system are necessarily individual living systems, but they are components only inasmuch as they modulate one another's parallelized states through their interactions in an operationally closed way. In contradistinction to self-referential systems, therefore, syn-referential systems do *not* modulate *the totality* of the states of their components, but only those states which participate in the formation of the social domain.

---

(1) I am indebted to my friend and colleague W.K. Köck for this term which seems to fit marvellously the idea I wish to express. Although the term "syn-referential" uses the same prefix as "Synergetic" (cf. the contribution of H. Haken in this volume), it has been chosen independently.

As it is, of course, not possible to explore the consequences of the proposed understanding of social systems (1) in great depth in the limited space of this contribution, I would like to mention just a few points of interest.

Social systems as syn-referential systems are defined - apart from other characteristics - by the operationally closed way in which their components interact. This
operational closure exists on two levels, on the individual level and on the social
level. As living systems, components operate in a closed way as has been explained
above. But the operational closure exists on the social level as well, and even in
two distinguishable forms. Every activity of a component of a social system modulates the other components. As this triggers a modification of their states, the
modulating component is modulated itself. Beside this inner feedback, there is
another which passes through the system's environment. This feedback is often more
diffuse in that component individuals of the system experience reactions of other
individuals to what these perceive as the system's behaviour. Whereas in the first
case a participant in the system often "knows" who started a given activity, this
is very often not the case when there is a feedback from outside the system. As
the proposed definition of social systems is based on individuals as components,
it is not difficult to explore the consequences of this twofold feedback. Leaving
this exploration to the reader's imagination, I shall just point at the consequences for *social change*.

Although social systems are conservative systems due to their organization, they
generate the phenomenon of social change. This can be explained as resulting from
the multi-component character of the individuals that constitute them. The inner
feedback of a social system is very often a conservative factor for the evolution
of a syn-referential system. In internally differentiated societies social change
seems to originate mostly from the interactions of social systems. Social systems
always interact through the interactions of their components, i.e. the individuals
that constitute the systems. As living systems they have to integrate their different existences as components of varying systems. If this becomes difficult for
whatever reason, they may modify their behaviour and the corresponding definition
of reality. This can cause a social system to change. It may as well happen that
components of one system have to interact with components of another system. In
this process the behaviour of the components of the respective other system constitute for the participants a description of the other system and its reality. If
there is no common social domain which can serve as a frame of reference for these
interactions, the descriptions can only be interpreted in the particular systemic
context. The result is a behaviour which at least is not coordinated, perhaps even
conflicting. This in turn leads to a modification of the interacting components and
hence to internal modifications of the interacting social systems. As this kind of
process happens all the time in internally differentiated societies, social systems
change and hence society as a network of interacting social systems. As these processes are brought about by the actual interactions of the social systems, social
change can be understood as a socially distributed process of the generation of
realities and of the adaptation to them, which leads to further processes of the
same kind.

References

Beer, S. (1975),   "Preface", in: H.R. Maturana/F.J. Varela 1975, 1 - 16
Benseler, F./Hejl, P.M./ Autopoiesis, Communication, and Society, The Theory of
Köck, W.K. (eds.) (1980), Autopoietic Systems in the Social Sciences,
                        Frankfurt/M., New York

(1) This has been done to some extent in P.M. Hejl 1982a, although the terminology
    used there, as well as the differences between the different types of systems,
    is not yet developed as far as in the present contribution.

Hejl, P.M. (1980),   "The Problem of a Scientific Description of Society",
                            in: F. Benseler/P.M. Hejl/W.K. Köck (eds.) 1980,
                            147 - 161

--- (1982a),   Sozialwissenschaft als Theorie selbstreferentieller
                            Systeme, Frankfurt/M., New York

--- (1982b),   "Die Theorie autopoietischer Systeme: Perspektiven für
                            die soziologische Systemtheorie", Rechtstheorie 13,
                            45 - 88

Hejl, P.M./Köck, W.K./   Wahrnehmung und Kommunikation, Frankfurt/M., Bern,
Roth, G. (eds.) (1978),   Las Vegas

Köck, W.K. (1980),   "Autopoiesis and Communication", in: F. Benseler/
                            P.M. Hejl/W.K. Köck (eds.) 1980, 87 - 112

--- (1981),   "On Communication and the Stability of Social Systems",
                            in: G. Roth/H. Schwegler 1981, 145 - 169

Luhmann, N. (1971),   "Politische Planung", in: id. 1971, Politische Planung,
                            Aufsätze zur Soziologie von Politik und Verwaltung,
                            Opladen

--- (1981),   Politische Theorie im Wohlfahrtsstaat, München, Wien

Maturana, H.R. (1970a),   Biology of Cognition, Rep. No. 9.0, Biological Computer
                            Laboratory, Dept. of Electr. Engin., Univ. of Illinois,
                            Urb. III

--- (1970b),   "Neurophysiology of Cognition", in: P. Garvin (ed.)
                            1970, Cognition: A Multiple View, New York, Washington,
                            3 - 23

--- (1974),   "Stratégies cognitives", in: E. Morin/M. Piatelli-
                            Palmarini (eds.) 1974, L'unité de l'homme, Paris,
                            418 - 442

--- (1978a),   "Biology of Language: The Epistemology of Reality", in:
                            G.A. Miller/E. Lenneberg (eds.) 1978, Psychology and
                            Biology of Language and Thought, Essays in Honour of
                            Eric Lenneberg, New York, San Francisco, London, 27 - 63

--- (1978b),   "Cognition", in: P.M. Hejl/W.K. Köck/G. Roth (eds.) 1978,
                            29 - 49

--- (1980),   "Man and Society", in: F. Benseler/P.M. Hejl/W.K. Köck
                            (eds.) 1980, 11 - 31

Maturana, H.R./   Autopoietic Systems, A Characterization of the Living
Varela F.J. (1975),   Organization, Biological Computer Lab. Rep. 9.4, Dept.
                            of Electr. Engin., Univ. of Illinois, Urb. III
                            (Reprinted in: id. 1979, Autopoiesis and Cognition,
                            Boston Studies in the Philosophy of Science, Boston)

Parsons, Talcott (1951),   The Social System, Glencoe

Parts and Wholes,   An inventory of present thinking, Documents from an
                           international workshop, arranged by the Committee for
                           Future Oriented Research in collaboration with Lund
                           University, June 1-3, 1983, Vol. 1, Swedish Council
                           for Planning and Coordination of Research, Committee
                           for Future Oriented Research, Stockholm 1983 (cited
                           as: Parts and Wholes 1983)

Roth, G. (1981),   "Biological Systems Theory and the Problem of Reduc-
                          tionism", in: G. Roth/H. Schwegler (eds.) 1981,
                          106 - 120

--- (1982),   "Conditions of Evolution and Adaptation in Organisms
                          as Autopoietic Systems", in: D. Mossakowski/G. Roth
                          (eds.) 1982, Environmental Adaptation and Evolution,
                          Stuttgart, New York, 37 - 48

--- (1984),   "Erkenntnistheoretische Probleme des Prinzips der
                          Selbstorganisation und der Selbstreferentialität",
                          forthcoming

Roth, G./Schwegler, H.   Self-organizing Systems, An interdisciplinary Approach,
(eds.) (1981),   Frankfurt/M., New York

Varela, F.J. (1979),   Principles of Biological Autonomy, New York, Oxford

--- (1981),   "Autonomy and Autopoiesis", in: G. Roth/H. Schwegler
                          (eds.) 1981, 14 - 23

Varela, F.J./   "Autopoiesis: The Organization of Living Systems, its
Maturana, H.R./   Characterization and a Model", Bio Systems, 5, 4,
Uribe, R. (1974),   187 - 196

von Foerster, H. (1981), Observing Systems, Seaside, Cal.

von Foerster, H./   Principles of Self-Organization: The Illinois Symposium
Zopf, G.W. (eds.)   on Theory and Technology of Self-Organizing Systems,
(1962),   London

Wake, D.B./Roth, G./   "On the Problem of Stasis in Organismal Evolution",
Wake, M.H. (1983),   Journal of theoretical Biology, 101, 211 - 224

Weber, M. (1976),   Soziologische Grundbegriffe, Tübingen

Yovits, M.C./   Self-Organizing Systems, London
Cameron, S. (eds.)
(1960)

Part II

Self-Organization and Management

Management – A Misunderstood Societal Function

H. Ulrich

Institut für Betriebswirtschaft, Hochschule für Wirtschafts- und
Sozialwissenschaften, Dufourstrasse 48
CH-9000 St. Gallen, Switzerland

1. "Management" and "Leadership"

If in writing this paper I prefer to use the term "management" rather than "leadership", it is not because of a predilection for modern words but rather to avoid any
misunderstanding. Leadership denotes in its most fundamental meaning "the leadership of people". When it is used in the context of heading an organization, the
concept of leadership has only a metaphorical meaning. The metaphor of leadership,
however, does not reflect adequately the complexities implied in leading social
systems. The result of using the term leadership when referring to heading   a
whole organization is that leadership of organizations is equated with leadership
of people *in* organizations. The term "management" by contrast does not convey such
a person oriented meaning. It is institutions that are managed, not people.

What follows is intended to show a) what management means from a system-oriented
theory of management, unencumbered by the conceptions contained in everyday language, and b) which societal functions thereby accrue.

2. Social Systems as Managerial Objects

Before management can be characterized as an aggregate of particular functions,
one must understand the true nature of the object to be managed because only when
related to that object do the functions of management become meaningful. In contrast to the majority of authors in management theory, whose writings relate to
private companies, we conceive the field of enquiry which must be of interest to
management theory in a much broader way: it includes all purposeful systems of
human society. This expanded view requires that such a system no longer be characterized from the standpoint of a concrete specific purpose, but rather on the
basis of an understanding of more general criteria, exhibited by all these systems;
this naturally dictates a more abstract perspective.

We perceive social systems as structures created by man in which people work together in order to satisfy societal functions. In reality, these functions can be
extraordinarily varied; generally speaking, however, they can be delineated as the
production of environmentally required services. In contrast to natural organismic
or ecological systems, social systems do not arise without human intervention; instead, they are the result of human purposes and actions. The precise point is
that specific goods and services are not produced by nature in the form and availability desired by society, with the consequence that societal systems are created. In this sense, these created systems are artificial, purposeful ones intended
to be responsive to human aims.

An important characteristic of such systems is that they are able to manifest themselves in their environment as entities, as authoritative and competent "persons".

For this reason, jurisprudence has among other things created the so-called "legal personality" where the term itself vividly denotes what is intended. An enterprise, for example, must be able, just as an individual, to assume rights and privileges even if it is not a "natural" person. On the other hand, it is obvious that systems are composed of a majority of "natural" persons, without whom they could not conduct their affairs. Nonetheless, a business partner of Nestlé Inc. would deem it inconsequential whether their contracting agent were named Smith or Jones. The important thing is that the agent represents the firm and can obligate it.

These social systems can be depicted from a systemic perspective as dynamic, open systems, i.e. as entities which 1) are circumscribable externally, yet evidence a relationship to the external environment and appear as a unit therein, but 2) are on the other hand, composed internally of recognizable components which are interrelated through a network of connections. Evolution theory has shown us that such systems may arise "by themselves" when previously unrelated elements, for whatever reason, enter into new relationships. It is important to recognize that such a new system will exhibit properties and behavioral patterns not derivable from the previously unconnected "parts" (cf. Lorenz, 1977); a completely new previously nonexistent entity will appear. One can also describe this process as the spontaneous creation or fulguration of a new order (cf. von Hayek's Theory of Complex Phenomena, 1967).

However, purposeful social systems do not originate by themselves but as a result of human aims and purposes. Although the spontaneous formation of new systems also plays an important role in society, social systems must be intentionally designed by human beings for human purposes. The goal is not some "self-generated" system behavior but rather one that is intended to fulfill the desired aims. From this point of view, social systems are "made" by human beings and are not nature-given systems.

As group dynamic shows, systems also spontaneously arise in society, which are termed "groups". Kurt Lewin has previously formulated as a basic proposition of group dynamics the principle that "A group is a dynamic whole" (Kurt Lewin, 1934). Additionally it has been established that groups can be perceived "as such" and that phenomena will occur which are not attributable to individual actions but rather those of the whole system. The creation of groups, however, rests upon the possibility of interaction between the participants, which presumes spatial proximity and unimpeded opportunities for communication. But it is exactly these prerequisites which are not "given" in social systems comprised of dozens to tens of thousands of participants. An old story typifying the lack of interaction between a large number of company employees is that of the two gentlemen who become acquainted somewhere and who only find out after a lengthy while that they are both employed by the same company. Social systems consequently are not groups in the sense of group dynamics but rather systems of another category, and neither their origination nor their behavior is explainable on the basis of group research results, just as little as one could explain group phenomena solely on the basis of the characteristics of the participants.

The visualization of social systems as "open" systems necessarily leads to the question of their relationships with their current environment. In this respect the analogy to natural dynamic systems, especially organismic and ecological systems, is once more closely allied. The natural sciences view the interactions between such systems and their environment as mutually adaptive processes which either lead to continuous "adjustment" of the system to its existing environment or its death. Once again, however, it is necessary to recognize that social systems do not exercise command over their adaptability because of a natural selection process. The ability to survive in a changing environment must to a degree be artificially created for those man-made structures intended to satisfy human purposes. In this case, moreover, it is not simply a matter of surviving "somehow"; the criterion for survivability is not a mere existence but the fulfillment of specific functions in human society.

The concept of open systems, from a systemic perspective, leads to a world picture that is composed of higher and lower order systems. Each system is then not only an entity encompassing multiple systems of a lower order but is also a component of a more inclusive system. Arthur Koestler uses the expression "holon" to indicate this dual nature and deduces therefrom, that open systems must simultaneously adhere to two contradictory tendencies: An integrative tendency, in order to function as a component of the higher order whole, and the self-assertive tendency, in order to uphold their own "autonomy" (Koestler, 1978).

In my opinion, the understanding of this dual nature of open systems is extremely important for the understanding of social systems. The ensuing consequence is that we must consider these open systems not only as systems but also as components of higher order systems. To a degree, then, social systems must be able to command two categories of attributes; on the one hand, whose which make them acceptable components in a superordinate system - the human society -, and on the other hand, those of an entity which is able in turn to integrate components into a higher order system. These two categories of attributes taken together constitute what one can describe as the "viability" of social systems. In this respect, the model of a viable system developed by Beer (1981) offers itself as a possible basic conception that is also useful in understanding social systems. But it must again be pointed out that a consideration of cultural accomplishments which social systems are, as if they were no more than nature-given systems, will lead to a reductionism inappropriate to the subject matter at hand.

The thoughts outlined here are in my opinion necessary for an understanding of what is meant by "management". Briefly stated, one can now conceive management to be that function designed to confer upon a social system those attributes that ensure its viability.

3. Management as the Design, Control and Development of Purposeful Social Systems

Management is mostly defined in the literature as the sum of explicitly tabulated activities such as planning, decision-making, organizing, leadership of people, controlling, etc. The deplorable aspect of such definitions is that numerous, varied lists of these activities are preferred by various authors and that a clear relationship between these individual activities is not established. In the following, we define management at a higher abstract level as the design, control and development of purposeful social systems.

By *design* is meant the construction of a system and its maintenance as a purposeful, competent entity. As previously indicated, systems do not originate by themselves but must be "brought to life" by humans for human purposes. The design task, therefore, is to select certain "people" and "things" from the environment and to convert them into components of a system exhibiting the desired viability attributes. The design task is by no means completed with the establishment of the system; the system's components must time and again be newly constructed and redesigned so that the system may be preserved as a whole. However, design as a management function does not consist of executing provisioning-type actions such as energy acquisition, material procurement and personnel recruitment. These concrete, executable activities at the operational level merely reflect the end results of management functions. Design as a management function denotes rather the mental envisioning of a system model which, for its part, necessitates the determination of the desired system attributes. These mental designs can be termed "design models", which can be clearly differentiated from scientific explanatory models (which aim to interpret an existing reality) and from decision-making models, which depict a specific problem situation in a given system. Design models in contrast, similarly to engineering drawings, portray a non-existent, yet-to-be-achieved reality. Their development consequently represents an eminently inventive process.

The immense difficulties which have to be overcome during the development of design models for social systems are connected with the extremely high complexity manifested by these systems and their environment. From this viewpoint, design can be considered to be the development of order which reduces to purposeful behavioral modes the potentially very great behavioral variety of an in itself complexly constituted existing system. At this point, the difference between the design of social systems and the design task of an engineer also becomes recognizable. In contrast to the design engineer's task, the development of social systems involves human beings as system elements, who are themselves viable systems exhibiting great behavioral variety and possessed of a value *per se*, which imposes human limitations on their reduction to carriers of particular system functions.

The reduction of possible behavioral manifestations to relatively few is bounded above all by environmental complexity and dynamics. The latter require that the system be capable of adjusting itself to a constantly changing environmental constellation, which presumes at any given time particular but not predictable behavioral modes. The demands upon a system's capability to produce in no time new behavioral modes which cannot be planned in advance will be all the greater, depending upon the degree of dynamism and discontinuous change shown by the environment. The indicated dual nature of open systems also finds expression in this fact. In order to safeguard its integral nature, the system must internally possess a considerable degree of order to restrict the behavioral variety of its components and elements. However, it must be in a position, as a component of a higher-order, larger system, to increase its own variety if necessary in adjusting to the superior system's behavior. If one sees management from this viewpoint as a process of overcoming complexity through a continuous reduction and generation of variety both in the entire system must always incorporate a considerable degree of uncertainty. Because future environmental constellations cannot be explicitly forecast it follows that a social system also cannot be designed to respond to the generation of specific, definable behavioral modes. On the other hand it is possible to conceive a system in such a way that, within delimitable *behavioral fields*, it has the ability to generate the then-required specific behavioral modes. The fleshing-out of these behavioral fields, in the end through specific, executive actions, must then be accomplished by means of another management process which we described as "control". An important design objective is that of providing the system with an attribute of "controllability".

We understand "control" as being the determination of goals and the establishment, execution and supervision of purposeful activities of a system or of its components and elements. Control is consequently a function which must be carried out in the system in order for the latter to achieve its purposes through specific actions. Control causes the system to select and realize very specific behavioral modes within a behavioral field defined and bounded by the system design. Because explicit transactioning cannot be effected by the system itself but rather only by its elements, control has as its ultimate objective that individual element that is to be generated in support of a particular system-serving behavior.

The concept of control corresponds best of all to the interpretations of leadership as they still prevail in the theory of entrepreneurial leadership. In this instance, separate phases or aspects of control processes are normally singled out. This becomes especially evident when leadership is defined in the co-worker leadership literature as an interactive relationship in which one participant (the leader), in order to achieve self-established goals, generates and maintains directed behavior by the other participant (the follower). Here we find the process of determining the behavior of individual system elements explicitly stated as being the subject matter of a theory of leadership. The decision-making-oriented theory of entrepreneurial leadership, by contrast, singles out a particular phase of the control process, namely the selection of a particular behavioral mode from the field, and establishes it as the subject of its investigations. Although both approaches, naturally, are legitimate and could contribute to a better understanding of certain aspects of management, they constitute a danger in that management could be understood as "nothing more than" leadership of people or decision-making.

The more abstract concept of control is, by comparison, naturally more comprehensive and was deliberately chosen so that the aspects of determining purposeful system behavior as developed by cybernetics could be included. If one begins with the classical tripartitioning of such processes into steering, regulation and adaptation, it becomes evident that leadership is predominantly reflected in the literature as a steering process in which the behavior of an element is determined by another, external one. Even if the perspective is expanded somewhat to include numerous elements in cooperation and having mutual interactions, the visualization of the situation as a process that can be isolated predominates over that of an open, partial system whose total behavior must be understood as adpation to the more inclusive system. The concept of a self-controlling, open system is therefore fundamental for an understanding of the control processes in social systems. The individual process of determining the behavior of an element assumes a meaning and is capable of meaningful execution only when it is embedded in this concept.

Hundreds of processes involving the direct control of elements take place daily in purposeful systems such as companies. They range from the initiation of preprogrammed machine activities through the selection of specific actions by the employee himself to the issuing of executive directives to the employees. The large number of dispositive control processes necessary to develop a "product", i.e., to achieve a goal desired by the system, is a consequence of the extreme division of labor and specialization within the work force. Consider, for example, a commercial airliner, which must not only depart on time but also arrive safely at its destination, must transport the right passengers and luggage, must serve the desired meals, and so forth. In order to accomplish this, countless services must be rendered by hundreds of people, - counter attendants, luggage carriers, airplane mechanics, meteorologists, airline controllers, pilots, hostesses, stewardesses, cooks, bus drivers, etc., all of whom must be selected, engaged and controlled. Or, if we consider a large firm in the machine industry, we can see that its products consist of hundreds of parts which are themselves mostly composed of various parts that are the product of thousands of operations performed partially within the firm and partially by so-called sub-contractors. Thus large firms have thousands of sub-contractors, whose activities must be integrated into those of the large firms and who, for their part, also have sub-constractors. As a result, a very complex framework of relationships is necessary both internal to as well as external to the firm, so that the desired end products actually can be achieved.

These examples are intended to show that the processes of directly controlling executional activities in such systems can only be completed in a time-wise and material-wise coordinated fashion when they are part of a more comprehensive *control system*. In this regard, it is to be noted that such control systems in today's industrial production firms are already built into the technical plants that are to carry out the executional operations. Automatons and automatic assembly lines represent to a large degree self-controlling operating systems which, under normal circumstances, make personal control measures by managers unnecessary. Where human actions predominate at the operational level, we find them increasingly being controlled by computerized production systems which, by means of automatic data processing determine and initiate the necessary executional actions. Anyone who visits a modern factory cannot help but notice this depersonalization of the direct controlling of executional operations; the work process itself rather than a superior determines the behavior of employees involved in the work. The visualization of leadership as an interactive process between superiors and subordinates thus completely fails to seize the essence of the control processes, as does the conception that it is a question of controlling a group. So-called managerial rules which are generally based upon this personnel-oriented idea of leadership, therefore, in my opinion, largely miss the point as to the true function of lower-level managers. This function consists of keeping the control system in check, immediate "hands on" intervention if malfunctions occur, settlement of exceptional cases, personally plugging gaps in the system and the linking of the system with others.

But it is also important that the management function, in the light of this development, has to a large degree shifted from that of direct and personal control to the *design of control systems*. Consequently in industrial operations production planning and control systems, material procurement systems, quality control systems, etc., must be developed, realized and kept under control.

The transformation of controlling activities into tasks of system design and supervision, indicated as occurring at the operations management level, is also increasingly taking place at the higher management levels. The tasks of middle and higher management are more or less abstract depending upon whether the design and control aspects concern smaller or larger subsystems of the whole system. The problem is not that of controlling the specific activities of individual elements but rather the establishment of those prerequisites which will ensure the successful execution of control actions. The representation of a multi-level chain of command presented as "official channels", which is still conveyed by organizational plans, is an exceedingly inadequate model for what actually happens in management. Thus, the important management function of planning cannot be properly understood if one considers it to be decisionary preliminaries which are the personal responsibility of an executive. This individual planning activity does indeed occur, but it can be meaningfully executed only within the framework of a planning system which encompasses it as well as hundreds of such individual activities and aims to determine not the behavior of a single element but rather that of the entire system or a relatively autonomous part thereof. In my view, it is extremely important to recognize that such higher order control systems possess a unique characteristic in that they do not relate to the activities of individual elements but rather to complex components of systems which are not only more, but are also in part characteristically different, than the sum of their parts. A primary task of higher management is therefore of a conceptional nature, i.e., it entails the development and establishment of systemic patterns which at a future point in time will effect the desired behavior by the whole system to be controlled. Because the foregoing, as previously indicated, involves the mental overcoming of high complexity and the precognition of future circumstances, the question arises as to whether there is any hope that such tasks can be mastered. According to the thought processes of rationalism (which, although coming under increasing criticism, is even today predominant in science and the practical world), it is reasonable to regard even such problems as solvable if sufficient logic and exact analysis is used. One result of this analytical thought process is the previously noted definition of management found in classical management theory which divides management into a set of individual activities which are then sequentially, and in isolation, further analysed. On the basis of this thought process, control systems are developed in the same manner as an engineer develops drawings for a new machine; these drawings must ultimately include all of the data required for the actual development of the machine. One can also say that the matter involves the establishment of a complete decision-making model or an aggregate of precise procedural instructions which need only to be followed in order that the machine will actually materialize. In an ideal case, then, the mental tasking involved in system development is completely accomplished beforehand by the engineer; this mental task only needs then to be realized during production in a material sense.

Although social systems are not machines and the creation of a "leadership machine" has yet to be successfully realized, many managers unconsciously accomplish their tasks on the basis of such a technocratic standpoint and are supported therein by a management literature that regards this mental perspective as the only scientific one. Symptomatic of this process are the efforts to completely mentally grasp a situation by means of always more exact measurement of situational behavior and always more information regarding particulars, in order to in fact control the situation.

In reality, however, the valid perception is the one formulated by Friedrich von Hayek as follows: "We have in fact learned enough in many fields to know that we cannot know all that we must know for a total interpretation of phenomena" (cf. Hayek, 1967). The science and practice of management have an extremely hard time with this recognition of a fundamental inability to know of many circumstances necessary to the exact control of complex situations. Instead, the philosophy of "still not enough knowledge" is untiringly championed and much effort is squandered to know things which are unknowable. As Hayek nicely shows, the functional efficiency of free enterprise is based directly on the fact that the total available knowledge is distributed in an extremely diffuse manner throughout the system and cannot be concentrated at a single point. In other words, the distribution of knowledge to all system elements is the necessary prerequisite for the creation of a self-controlling system, whereas, contrariwise, the attempt to centrally control the economy is already doomed to failure because of the "inability to know", in the sense that individual controlling agencies do not have at their disposal the necessary knowledge.

Such visualizations of the functioning of social macro-systems can also be applied in their essential features to individual purposeful systems, as these also exhibit the attributes of high complexity. The only tenable basic conception for the design of social systems thereby becomes the model of the self-controlling system.

From this viewpoint, management's task consists above all in providing social systems with a capability for self-control. Such systems exist in nature in the form of organisms and ecological systems. One can therefore envision the mentioned model of a viable system as being a self-controlling one and by analogy to natural systems, attempt to apply it to social systems. Doing so may involve, on the one hand, understanding those attributes and functional modes which provide the system with its self-controlling capability or, on the other hand, posing the question as to how such systems arise in nature in the hope of determining principles for the deliberate design of social systems.

Evolution theory teaches us that natural, viable systems do not arise because of a technocratic approach but rather by process of mutation and selection occurring over extended periods of time, i.e. through changes in the living state occasioned by chance, which result from "small errors" during the reproduction of genes, as well as from processes of adaptation to the environment. "Evolutionary management" on a cultural level cannot, however, consist of simply letting this time-consuming evolutionary process, whose path is paved with the corpses of non-surviving systems, take place. It is possible, nonetheless, to adopt the conception, based on an "evolutionary approach" that complex viable systems are not "makeable" according to a plan developed in advance, but rather can only be developed over a passage of time (cf. Malik/Probst, 1982). This leads to the conception that the control and design of social systems must be understood as activities occurring within the framework of a long-term and never-completed system *development process.*

We also see now that these developmental processes cannot be left to their own devices but instead must be deliberately designed and controlled. The survival of a social system and its continuous functional performance should not, after all, be entrusted to the chance happenings of a natural evolutionary process. The concept of business and management development worked out at the St. Galler Management center presents a design model for such a "guided development" (cf. Malik 1981, pp. 49). In this model the control of corporate development occurs from two perspectives: firstly, through the periodic formulation of conceptional corporate images on the basis of previous developments and current corporate and environmental circumstances, and secondly, through continuous controlling measures based on such concepts. The concept of a self-controlling system fits perfectly well within this model, in that these conceptional design and control measures need not be developed and prescribed by superior authority, but instead are to be developed and realized by those personally participating in the development process.

This kind of active and continuous human co-operation in such developmental processes can only take place within small groups in which the previously mentioned opportunities for direct communication are present. In order to really wring out problem solutions it is therefore necessary, in larger systems, to have numerous teams of this type, which together constitute a hierarchical, interlinked ad-hoc structure incorporating numerous feed-back loops.

If one accepts this type of institutionalized, continuous further development of the system as an additional management task, then the picture of middle and higher management's role is as follows: their actions as direct activity controllers are increasingly confined to dispositive decisions in so-called exceptional cases and to intervention in the case of malfunctions, not on the basis of the classical principle of delegation, whereby the authority to resolve "normal" situations is to be delegated by superiors to the subordinate, individual co-workers, but rather because the system itself commands self-controlling operational and dispositive systems. By comparison, the design task moves into the foreground for these managers, requiring the design, activation and control of such self-controlling systems for the existing next-lower system level. A third task is increasingly being added, namely assisting in the continuous, long-range development of the entire system in the manner described.

In summary we can say that control, design and development represent the constituent functions of management which are distinguished primarily through the differing range (both temporal and as far as substance is concerned) of their intended effects. If we use the currently often-applied dual division into operational and strategic management as a point of departure, the control function (on the basis of the centrality of its activities) must be assigned to operational management and the development function to strategic management. The design function, which cannot be accommodated by this simple division,then consists of designing operational systems in such a manner that they are controllable within the framework of the behavioral fields defined by strategic management. If we start with the visualization of a multi-level hierarchy of managers, and differentiate between lower, middle and higher management levels, then the three management functions can be depicted in terms of their level-oriented weighting as shown in Figure 1.

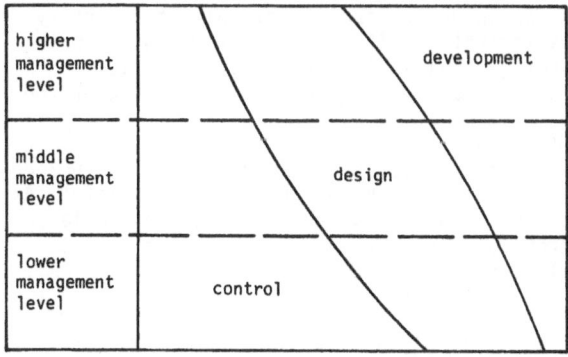

Fig. 1:
Management levels and functions

4. The Dimensions of Management

Social systems have been frequently compared, above, with natural systems, but the special characteristics of systems developed by humans to satisfy human purposes have also been pointed out repeatedly. Despite these references an impression may have been created that the human being in this instance is once more being left out of consideration. The following is intended to show that exactly a rather formal depiction of the management function will provide a perspective for an understanding of the human aspects of designing, controlling and developing social systems.

In order to understand social systems, I consider it expedient to differentiate between three levels, which can briefly be described as material, biological and cultural (Fig. 2), and which correspond in their typical viewpoints to those of the three scientific groupings of physics and chemistry, biology and ecology and cultural sciences.

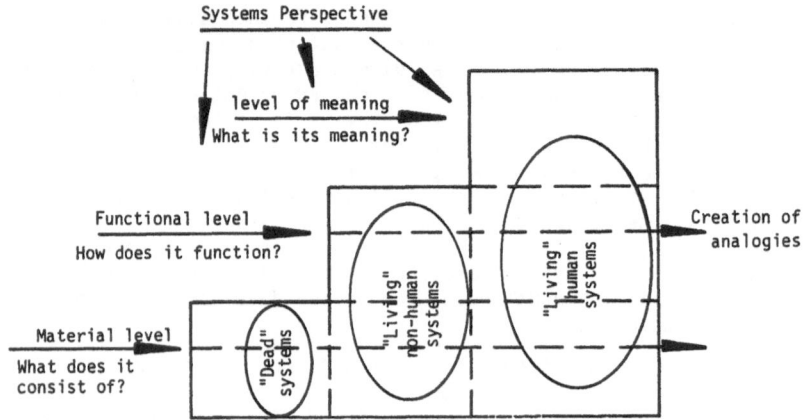

Fig. 2: Perspectives, Dimensions and Objects of Knowing

Social systems can be understood on the lowest level as systems that absorb convert and discharge environmental matter and energy. These process forms occur continuously, not only in natural systems but also in social systems; this is naturally most clearly evident in the case of those industrial activities producing material goods. It would be completely impossible to make them controllable systems serving human purposes if they were not also designed at this level. J. Rifkin (1980) shows very impressively in his comprehensive view of the consequences of the law of entropy that such a viewpoint can be fruitful for an understanding not only of "dead" systems but also of the processes in viable systems. It would be wrong to evaluate the usefulness of this material perspective against the result if one tries to view cultural phenomena in the light of a nineteenth-century physical view of the world.

The perspective from the intermediate level orients itself according to the perceptions of those sciences which concern themselves with the phenomenon of "life". As indicated more than once, this perspective leads to the visualization of social systems as viable systems and the search for design principles corresponding to those in the structures and behavioral patterns of organic or ecological systems.

The current numerous attempts to develop an "evolutionary theory of management" are largely assignable to this intermediate level. It should also be noted here,

that the usefulness of such a conception does not lie in a return to a classical
and naive "social Darwinism" or the use of statements from classical biology,
which in one philosopher's opinion, has understood how "to reduce the living to
the dead" (Löw, 1980, pp.12), but rather in the still not fully exploited knowledge potential of the "new biology" and ecology.

Finally, the perspective from the third level includes those characteristics attributed only to humans and social phenomena. System processes as seen from this level
are not natural occurrences which are capable of being traced back to their causes
and whose results are predictable, but events which become meaningful for human
beings. The system, in the absence of an assignment of values to conditions and
events will remain incomprehensible and meaningless and rational human actions as
well as purposeful system behavior will not be possible.

This level can be described as that of the humanities, liberal arts or cultural
sciences. Its recognizable distinguishing features are constitutive for social systems and would not exist without human intentions and purposes, judgements, pro-
cedual motivation and behavioral norms. Although a humanities-oriented consideration of social systems has indeed been accorded greater significance since the
human relations movement of the 1930's, it is still closely bound to an individualistic psychological conception. From a systems perspective, however, the system
presents itself within this level as a complex, dynamic and network-like entity of
human values and norms, whereby individual behavior in this system cannot be understood on the basis of individual motivation but only within the context of
the entire norm structure constituting the system.

Management can now be understood as a sense-making process through which human
meaning is imparted to circumstances and events. In this light, the controlling
function consists above all in assessing existing or expected circumstances,rating
them as either "good" or "bad" and deriving therefore the desired actions that
would lead to a "good" state of affairs. When managers are termed "decision-
makers" or "problem-solvers", it is misleading in that the associated concept is
one that "sees" problems as "given" and as something that needs merely to be understood and solved. But problems are the products of human judgements; they cannot
be found like an object, but must instead be contrived. This problematisation of
actual or anticipated circumstances and processes represents a primary management
task.

Management's design function on this intellectual level consists in devising a
system of normative rules that will guide system behavior towards values and aims.
In contrast to controlling, design does not consist in establishing individual
norms for behavior in current situations, but rather in creating a meaningful order of more abstract norms which, above all, should inhibit undesired behavior; in
other words, the determination of behavioral fields in the previously indicated
sense. The development function as viewed within this intellectual dimension consists in ensuring that these sense-making processes can continuously and successfully take place.

These deliberations lead to the conclusion that management is a profoundly human
function which can be accomplished only by human beings. Systems need managers in
order that this sense-making function will take place or, put another way, co-workers explicitly charged with this sense-making function, are described by us as managers. This does not mean that their personal values are decisive in their decisions and actions, i.e. that in the final analysis, everything would be in this
sense the result of "subjective" judgement. Managers can act sensibly only as elements of a system-related supra-individualistic value system in which they serve
concurrently as exponents and co-designers. To consider leadership as a negotiating
process involving the subjective "interests" of participants is therefore neither

an appropriate nor tenable proposition. The "interests" of the institution are not derivable from the "interests" of its elements but only through a comprehension of the system as an autonomous, environmentally dependent entity.

However, the attempt to "objectivize" the management function in the sense of eliminating human values is also equally senseless. The classical postulate of value free action and judgement and an unreflective application of natural science research principles to the investigation of cultural phenomena have unfortunately additionally misled social scientists into becoming enamored of such efforts at "objectivization". But such a reductionistic perspective is also prevalent in the practices of our scientific-technological area. This perspective distinguished by the negative connotation often associated with the expression "subjective", and the high esteem given to "objective data", even if they only consist of statistical averages, e.g., of "public opinion". Also arising out of the foregoing is the attempt to justify decisions by the application of exact, logical methods, although it must be obvious that human actions cannot be substantiated on the basis of selecting the correct, logical alternative but only because of the purpose and effects of those actions. Objectivization within the sphere of human actions can only sensibly mean that value judgements are not arbitrarily and thoughtlessly accepted as subjective opinions but rather because of their meaning as tested in the light of a higher order value system. The individual responsibility of managers lies in this necessity to justify personal values and decisions with the help of a higher order value system.

It also thereby becomes evident that management inescapably presents an ethical aspect. Decisions concerning what the system is to be and what it is to achieve can be based only on decisionary criteria stamming from what "should be"; only then can the responsibility for management be borne. This next higher level of norms for a social system requiring the latter's orientation with respect to its desires and aims is the level of a social value system and "there is no economic funnel that will reduce social value dimensions to a single economic dimension" (P. Ulrich, 1981).

If one views these three management dimensions, it becomes apparent that management functions affect all three levels, and, indeed, in one important aspect, the creation of the connections between them. Management is in its nature an intellectual, judgemental and sense-making function, but it relates to systems which have the task of making such values available to other humans by means of their conversion to a material form. It is true that design, control and development are primarily mental projections of possible situations and actions, but they only become meaningful if they create an effect in activities and organizations on a material level.

This necessity, the conversion of thought to action, to see that ideas become material states, sets the manager apart from a mere thinker who does not consider himself responsible and accountable for the realization of his thoughts. It is this span from "thinker" to "maker" that occasions the many-sided attributes and capabilities necessary for successful leadership.

5. Management as a Social Function

Today's society can without exaggeration be described as highly organized. Hundreds of thousands of systems have been created in a historically short period of time in order to satisfy an unlimited number of differing ends. Although humans have always understood how to create these structures in order to reach goals not attainable through individual actions alone, the multiplicity and density of today's systems is, nevertheless, historically unique. Society, today, is therefore to be less regarded as a sum of humans but more as an extremely complex system in which the institutions exist as components. A characteristic of the modern, institutionalized society is that it is impossible to develop a logical category system for the classification of all social systems, as everyone knows who at some time has tried to establish an ordered enumeration of systems active in only one sector of society.

All these so materially different systems, be they now private enterprises, public services, state administrations, hospitals or schools, have in common that they must be designed, controlled and developed if they are to fulfill their function in society for a long period. Management consequently has become a decisively important function for the operations of an institutionalized society.

One can hardly now assert that today's institutionalized society is performing particularly well; on the contrary more and more undesirable and undesired situations seem to be becoming more frequent in pratically all fields of life. Much that yesterday was still functioning faultlessly, today periodically gets out of control or produces negative side-effects in no longer acceptable proportions. It often appears that systems have completely divorced themselves from the supervision of their supposed controllers and are operating according to self-generated norms that no one wants. This suggests that the management of the system obviously is not as good as it should be from a social perspective, which, on the other hand, in my opinion, results from the fact that the true meaning of management has up to now not been widely understood.

The fact that the overall management function in larger systems is exercised by hundreds or thousands of managers and that only these individual management activities are concretely visible hinders a comprehension of the function as a whole. The prevailing analytical and isolationary thought processes also lead to a recognition of these individual activities as vital ones and to the subsequent endeavors to recruit employees with the corresponding capabilities as managers. Even management theory is to a large extent nothing more than a systematized knowledge of individual management tools and techniques.

The ideas presented in this paper aim to show that the essentials of management cannot be comprehended within this narrow viewpoint and that a wider perspective is required. I recognize that such a form of abstract presentation appears alienating and (in deterring sense) theoretical for those very people who are daily engaged in concrete management tasks. It nevertheless appears necessary to me, if one wants to understand the management world, to choose the broadest possible perspective, in order to see one's own partial functions within a larger context; it is also necessary to adopt a more abstract way of thinking, which will allow the not physically viewable, intellectual aspect of this social function to emerge.

The viewpoint advocated in this paper sees management relating itself to systems and their components which are not understandable simply as masses of individuals but only as inter-related entities which, based on their order pattern, develop their own behavioral modes. Because such systems and their structures are not visible and are immediately destroyed in an analytical thought process, their intellectual comprehension presupposes an abstract and holistic thought process which we were not taught in school. Because of the displacement of the focal points of the management function itself, such a form of thinking has become urgently necessary for higher level managers in as much as their task has increasingly been shifted from directly controlling individual activities to the design and development of entire systems.

There are, indeed, many systems which perform quite well for a period of time and in accordance with the top managers' norms. Should unexpected greater difficulties arise, however, and the charge of "management mistakes" then logically be placed at their doorsteps, they indignantly point to the unpredictable changes in the environment which caused the failure. In this case, the alibi is better than the charge, because drastic environmental changes truly cannot be predicted and the systems' adaptability is confined within narrow limits by all sorts of "situational constraints". But this only means that we have not yet learned how to design systems so that they will be viable in the face of the anticipated turbulent environmental changes of the future.

It must justly be said, however, that the individual systems are not free to autonomously determine their behavior because within the larger so-called macro-system they are merely an element, which must subordinate itself to the prevailing rules of this higher-order system. These abstract systems of a higher order such as the law, the economy, public health, etc.,have developed into monstrous systems, that are neither intellectually nor practically capable of being controlled by anybody. These systems, however, are also not entrusted to the powers of self-organization and self-regulation; instead, the politicians consider themselves at best as steering "organs", most often, however, as controllers of individual activities within such systems. Politicians exercise management functions at the highest levels in society as it is their task to judge, and, if necessary, to design these large social systems and to provide for their sensible development. But because they do not normally interpret their task in this fashion, but still consider themselves in accordance with old leadership precepts as helmsmen and actual controllers, they literally do not know what they are doing, because what they must know in order to intervene sensibly in complex system relationships belongs to the category of the basically unknowable. But here, too, it must be added in all fairness that we have not been able to learn anywhere how such historically new, giant, extremely complex networks can be intellectually grasped, sensibly designed and successfully controlled towards human ends. We lack experience as well as the requisite intellectual tools.

The new discoveries in the natural sciences will no doubt enable us to understand better and better how natural systems are able to survive, and, with an increasing measure of success, to try to develop models, based on such findings, of viable social systems. As our consideration of the three dimensions of management shows, such a form of functional thinking alone will not suffice. The core of the management function continues to be the understanding of social systems as cultural phenomena and, with that, a conscious confrontation with the question of human value judgements, and the harmonizing of one's own desires and aims with those of a higher order value system, which can only be perceived and understood by viewing one's own system from a societal viewpoint.

References

Beer, S., (1981)   Brain of the Firm, 2nd Ed., Chichester

v. Hayek, F.A., (1967)   The Theory of Complex Phenomena, in: Studies in Philo-
                            sophy, Politics and Economics, Chicago; also published
                            in M. Bunge (Ed.), The Critical Approach to Science and
                            Philosophy, Essays in Honour of Karl R. Popper, New
                            York 1963

Koestler, A., (1978)   Janus - A Summing Up, London

Lorenz, K., (1977)   Behind the Mirror, A Search for a Natural History of
                            Human Knowledge, London

Löw, R., (1980)   Philosophie des Lebendigen, Frankfurt

Malik, F., (1981)   Management-Systeme, Volksbank-Broschüre,
Die Orientierung, Bern

Malik, F./Probst, G.,   Evolutionary Management, in: Cybernetics and Systems,
(1982)   Int. Journal, 13, 153-174; also published in this
volume

Neisser, U., (1976)   Cognition and Reality, San Francisco

Rifkin, J., (1980)   Entropy - A New World View, Viking Press N.Y.

Ulrich, H., (1981)   Die Betriebswirtschaftslehre als anwendungsorientierte
Sozialwissenschaft, in: N. Geist, R. Köhler (Ed.),
Die Führung des Betriebes, Stuttgart

Ulrich, H., (1968)   Die Unternehmung als produktives soziales System, Bern

Ulrich, H., (1982)   Anwendungsorientierte Wissenschaft, in: Die Unternehmung, 1, 1-8

Ulrich, P., (1981)   Wirtschaftsethik und Unternehmungsverfassung: Das Prin-
zip des unternehmungspolitischen Dialogs, in: H. Ulrich
(Ed.), Management-Philosophie für die Zukunft, Bern,
57-

Weick, K.E., (1979)   The Social Psychology of Organizing, 2nd ed., Reading
Mass.

Systems Thinking in Management: The Development of Soft Systems Methodology and Its Implications for Social Science

P.B. Checkland

Department of Systems, University of Lancaster, Lancaster, LA4 1YX, England

Introduction

A decade of action research has produced a systems-based methodology for tackling the messy ill-structured problems of the real world, problems so much less tidy than those of the natural scientist in his laboratory. The methodolgy has been used at least two hundred times in problem situations of different kinds in both user-supported and public-supported organisations. It has shown itself to be both teachable and transferable to other users.

If we take it that the methodology, known as "Soft Systems Methodology" (SSM), has been shown to be "useful", then its implications for our view of "social systems", "management", and "self-organising systems" are worth examining carefully in the light of that success.

This paper offers a savagely brief account of SSM and the systems thinking neces-sary during the course of its development, together with a discussion of its im-plications for the nature of social reality and hence for our perceptions of "human systems". Briefly described are: an intellectual context for the work; the histori-cal context in which SSM was developed; the systems thinking developed during the emergence of SSM; the nature of SSM; its implications for our views of social rea-lity.

A full account of SSM's development is given in Checkland (1981b), which also has an extensive bibliography.

An Intellectual Context

It will be helpful to declare an intellectual context in which this paper (and the work it describes) make sense. This will then be used at the end of the paper to structure the discussion.

I take as given the existence of a real world of great beauty, squalor and comple-xity. We are, as human animals, part of that complexity; but by a familiar (but mysterious) mental act we can perceive ourselves perceiving that real world, as in Figure 1. We can try to make sense of the world by creating intellectual construc-tions x1 x2 x3 ... and trying to map them onto the world's complexity. Thus we have reality R, our intellectual constructions x, etc. and a methodology M in which we map x onto R. Of course we are here already the prisoners of a Foesterian "2nd Order" complexity: the source of many of our x's will be R itself, so that we have, not R and M-using-x as separate entities, but a basic complementarity. R is the source of many x's which are themselves used in M to enable us to perceive R.

Figure 1: The Intellectual Context
of this Paper

The position I take is that systems thinking is simply "one of the x's". Intuitive notions of R as densely interconnected - more like a hedge than a handful of marbles - lead us to develop formal concepts of "systemic wholes" which we then try to use to understand R better.

The work to be summarised here was a conscious attempt to use systems ideas in tackling one problematical aspect of R, namely the situations which strike us as "problems" and call for our purposeful intervention. The focus was thus on the kind of real problem situation which institutions have to deal with, the kind of problem situation with which policy makers or managers at any level must deal.

The Historical Context

SSM was developed by a university department of systems which was founded as a department of systems *engineering* (Jenkins 1967; Checkland 1970). Initial skills were in statistics, control engineering and chemical engineering; but it was always the intention to read the word "engineering" in its broad sense, and SSM grew out of a project to ascertain what happened to the well-established methodologies for engineering systems to achieve defined objectives (Hall 1962; Chestnut 1965, 1967) when this type of methodology was applied to "soft" problems such as those of policy makers, administrators, managers.

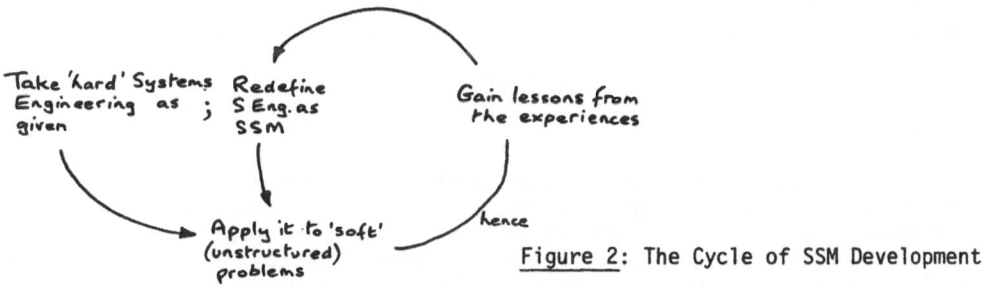

Figure 2: The Cycle of SSM Development

The process was one in which systems engineering was applied to unsuitably "soft" problems in organisations, and was found wanting. Lessons from this then served to redefine the starting methodology. "Systems Engineering" became "Soft Systems Methodology" as a result of the cyclic experiential learning. What was difficult to grasp was the *extent* of the learning garnered in the cycle of Figure 2. Many experiences and several years of reflection went into understanding how markedly SSM differed from systems engineering. To gain that understanding it was necessary to do some very basic systems thinking, which is summarised in the next section (Checkland 1983, gives a more detailed summary).

Relevant Systems Thinking

Typology 1

Among the x's of Figure 1 which comprise systems ideas are some which have been very successfully mapped onto aspects of the real-world complexity R. It has been possible to make systems models in various forms which map real-world physical wholes which are themselves *natural*. For example, various aspects of the physical world, and of fishes, flowers and forests upon or within that world, can be eluci- dated by the use of system models. (Indeed, so successful has such mapping been that in everyday language we slip into the category mistake of referring to the real-world entities as "natural systems". We forget that "systems" are some of the x's of Figure 1; they reside in the minds of human observers, and strictly spea- king we are wrong to talk of the real-world wholes as *being* "natural systems". Technically, pedantically, we should describe them as "natural entities which can be successfully mapped by system models". This mistake is usually made whenever mapping is successful; and is usually unimportant in everyday discourse. But we should note it carefully when developing systems thinking fundamentally.)

Natural physical wholes are not the only kind which can be mapped by system models. Our world contains many physical wholes which are not natural but are *designed*. They too can be mapped in various aspects by system models. The systemic models of the systems engineer for example are useful both in creating and managing chemical plants (where, again, in everyday discourse, we slip into referring to the plant as *being* a designed physical system). Other useful systems models map real-world wholes which do not have a physical referent in the world. For example, the desig- ned abstract systems of cybernetics include some - we refer to them as "control systems" - which map the connected processes by means of which some wholes, whether natural or designed, can retain their integrity within environments which change and impose new demands upon the entity. Also we may note that *any* systems model is itself an example of a "designed abstract system".

Thus, among the x's of Figure 1 are concepts of natural systems, designed (physi- cal) systems and designed (abstract) systems which can be used to map onto some aspects of the real world R. The development of SSM was a project to use systems concepts in tackling that part of R which can be described as "unstructured real- world problems" and in so doing either use an existing M - 'hard' systems enginee- ring - or develop an alternative. Events and experience forced the adoption of the latter course; and in taking it, it was found necessary to extend the systems typology described above.

Real-world problem situations often entail entities mappable by the concepts of natural or designed systems. But most intensely they entail human beings engaged in *purposeful action*. It was found extremely useful to develop the system concept of "a system which takes purposeful action", a system type called a "human activi- ty system" (HAS). It was envisaged that such an 'x' would be useful in trying to understand purposeful activity in the world.

It was quickly discovered that systemic concepts of human activity could not be described (technically, systems models of HAS could not be built) unless an ex- plicit *Weltanschauung* were defined. A system model of purposeful activity which is to be relevant to a study of real-world prisons, for example, has to be built ac- cording to a particular *Weltanschauung* with regard to prisons: rehabilitation, re- education, punishment, protection of society, etc. To examine real-world purpose- ful activity requires *clusters* of HAS's embodying different points of view, each being, explicitly, a formally constructed (systemic) Weberian "ideal type", not a would-be model *of* the real-world purposeful action (Checkland 1981b, pg. 270).

96

Because of this characteristic property (a) that HAS models require an account of both the model and the *Weltanschauung* which makes that particular model meaningful, and (b) that alternative *Weltanschauungen* will always be plausible, we give up the idea of a single system model *of* any purposeful activity (real-world prisons are complex, changing mixtures different for different observers, of rehabilitation, re-education, punishment, the unintended training of criminals,etc.,etc.) and replace it with the notion of a cluster of models *relevant to* the real-world manifestation of the activity in question.

This had a longer-term outcome for SSM which may be previewed here. It uses systems models (its chosen x's in Figure 1) but in a way which transfers the notion of systemicity from the world (R) to the process of enquiry into the world (M). The idea that R *is* systemic is a defining characteristic of 'hard' systems engineering which did not survive the attempt to apply systems engineering in "management" problems, broadly defined.

Typology II

The systems thinking outlined above focuses on what types of system model might be useful in enquiring into the world's complexity. During the development of SSM, its application in problem situations of different scope in different types of organisation, and comparison with other model-based approaches within management science, drew attention to the possible classification of the types of situation which arise in, or are perceived in the world and are deemed problematical (Checkland 1981a).

Because the successful intellectual and practical adventure which is natural science depends upon a regular patterned, rather than a capricious universe, one may postulate that there will be in the world situations or phenomena characterised by interconnections which are part of *the regularities of the universe*. Examples would be such natural wholes as mice and mushrooms, or the biosphere of planet earth. Call these phenomena of Type 1.

Secondly there will be situations characterised by interconnections which derive from *the logic of the situations*. It is not chance which leads to the fact that the same mathematical equation covers the growth of the scientific literature and the loss of weight of a starving rat. Here, in the case of the growth of the scientific literature, situational logic dominates, as it does in the problem types beloved by management scientists (the queueing situation, the depot location problem,etc.). Call these phenomena, or situations, Type 2.

Thirdly, many situations do not exhibit natural regularities or situational logic. They are dominated by *the meanings attributed to their perceptions by autonomous observers*. These Type 3 situations are ubiquitous in the world of human activity - e.g. how should we behave towards ageing parents? Should the nuclear deterrent be abandoned? What should we do about the penal system?

The original 'hard' systems approach (systems engineering, RAND systems analysis, OR) was developed in and is suitable for use in situations of Types 1 and 2. Natural science deals with phenomena of Type 1. SSM found itself trying to cope with Type 3 situations dominated by human perceptions. There, the most useful of the systems 'x's were the clusters of HAS models relevant to a problem situation; and in general there is likely to be a relationship of the following kind between the two typologies:

Natural system models	will map	Type 1 Phenomena
Designed system models (Physical and Abstract)	will map or be most relevant to	Type 2 Situations
HAS models	will be needed to elucidate	Type 3 Situations

The Nature of Soft Systems Methodology

SSM is a cyclic learning system which uses models of human activity systems to explore with the actors in a real-world problem situation their perceptions of that situation and their readiness to decide upon purposeful action which accommodates different actors' perceptions, judgements and values (what Vickers, 1965, calls their "appreciations").

The brief account of the methodology given here will describe a sequence of steps which constitute the cyclic process; in practice, sophisticated users can enter the cycle at any point, and, indeed, move temporarily *against* the sequence of activities described here. What matters is the nature of the stages making up the cycle and their mutual relations rather than the precise sequence in which the activities are carried out. (That this is so is the result of the basic complementarity which characterises SSM: real-world perceptions and actions enable "relevant" HAS's to be selected; relevant HAS's enable the real world to be perceived in a way which leads to purposeful action, as in Figure 3.)

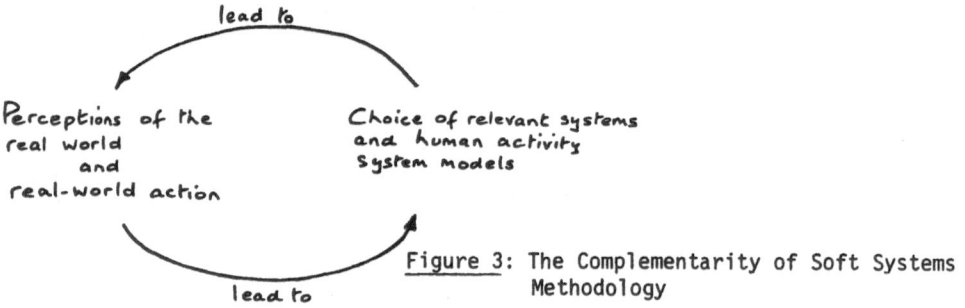

Figure 3: The Complementarity of Soft Systems Methodology

The form of SSM is shown in Figure 4. In Stages 1 and 2 a situation in the everyday world which is regarded as problematical is tentatively "expressed". This is done "tentatively" because the methodology has learned never to try sharply to define "the problem". Such definitions always imply a leap to "solutions", and so eliminate the learning, the change of appreciations which it is the main purpose of SSM to orchestrate. All the experience of developing and using SSM runs counter

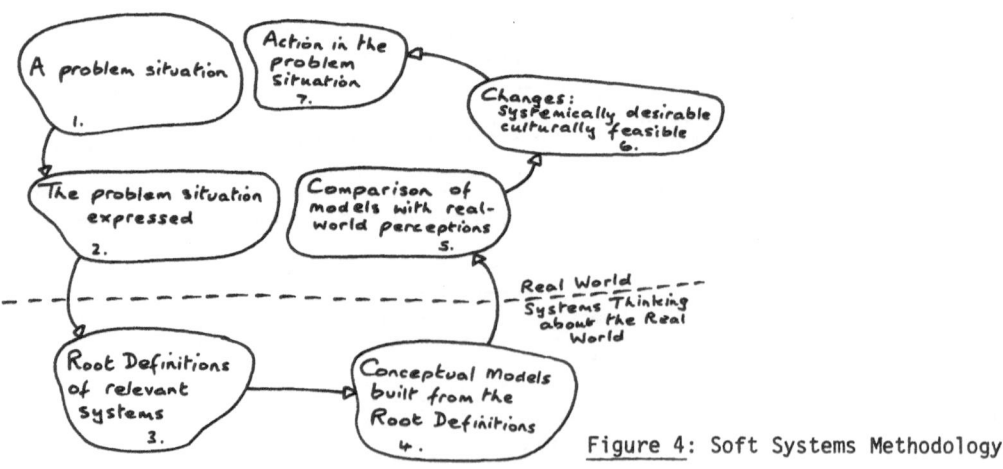

Figure 4: Soft Systems Methodology

to the main thrust of that considerable literature on "problem solving" which is
predicated on the proposition that problems may be, and should be defined as a
search for an efficient means to achieve a defined objective (e.g. Newell and
Simon, 1972). The impossibility of defining objectives once and for all and the
conflict endemic in different actors' definitions is characteristic of unstructured real-world problems.

It is important not immediately to perceive the problem situation systemically,
when "expressing" it, since this tends to cut off radical channels of thought. The
aim of Stages 1 and 2 is to obtain sufficient understanding of the problem situation to name some HAS's relevant to exploring it. Useful guidelines here are to
list elements of "structure" and elements of "process" in the situation and then
to record the nature of the interaction between the two.

Now begins, in Stage 3, the formally organised systems thinking about the problem
situation which is completed in Stage 4. In Stage 3 HAS's deemed relevant are carefully named in so-called root definitions (RDs) which express pure *Weltanschauungen* - for example, root definitions based separately on rehabilitation, protection of society, punishment of law breakers and the training of criminals might
all be formulated in a study concerned with prison management. Technique to help
in the proper formulation of RDs is available but is ignored here; see Checkland
1981b, Smyth and Checkland, 1976, and Checkland and Wilson, 1980. In Stage 4 the
definitions from Stage 3 are used as the basis for building models of the systems
of purposeful activity named in the RD's. The modelling language is simply "all
the verbs in English", and model building consists of structuring the minimum necessary activities required in a system which meets the requirements of the RD.
Well-formulated RDs help model building, of course, and there is also available
an experientially derived general model of any purposeful activity system. This is
helpful in ensuring that the models are complete and well-structured (see Checkland 1979, 1981b).

With the completion of the model building, which is best done by - or at least
with - actors in the problem situation, we are ready to expand exploration of the
problematical situation by using the system models in a comparison with the real
situation. The models are not normative; they are not designs; their purpose is
to orchestrate a debate about the problem situation. Frequently this Stage 5 comparison (for which four different techniques have been used - see Checkland 1981b)
leads to new choices of relevant systems, sometimes ones which open up new aspects,
sometimes ones expressing *Weltanschauungen* counter to those originally entertained.

The comparison stage debate ends when it is possible (Stage 6) to formulate some
changes which achieve *accommodation* (not necessarily consensus) between the different interests, perceptions, values and ideologies which characterise the problem
situation. Stage 6 seeks changes which are simultaneously desirable according to
the systems thinking in Stages 3 and 4 and *culturally feasible* for these actors
in this situation.

Action to implement the changes constitutes Stage 7; this changes the situation
itself, and the *new* problem situation may now be tackled: the cyclic learning process can begin again.

An Illustration

The illustration which follows is not to be taken as a full account of a systems
study using SSM - such a task would require the equivalent of a novel sequence,
preferably one written by a consortium of authors! What follows is intended as an
outline intended to put some flesh on the skeleton of abstraction just described.

Consider the complex charity organisation Oxfam. It owes its name to its origins in the *Oxford* Committee for *Famine* Relief. It has grown vigorously from these origins. Money-raising Oxfam shops exist in many British towns selling second-hand goods and Third World products. Oxfam engineers may be observed installing water pumps in villages in India. Newspaper advertisements invite donations. This is a complex of activity managed in many cases by people who have left better-paid jobs in industry to join Oxfam.

Consider a systems study to help improve the management of the organisation. Building a rich picture of the problem situation entails reading documents and interviewing both people in the organisation and people who have connections with it. A problem theme emerges: the flow of information to a regular monthly meeting of the management committee at the Oxford headquarters.

It might be useful, as the formal systems thinking starts, to have available some systems models relevant to (not *of*) Oxfam as a whole. Root definitions of some relevant systems are formulated based on: a relief system; an aid-provision system; a retail system; a system to persuade the rich world to devote more of its resources to the Third World ... These definitions are, in SSM, the general case (of unallocated perceptions) which, in 'hard' systems engineering, becomes the special case of (allocated) "defined objectives". The activity models from detailed definitions derived from these core ideas are now built. *They are not an account of Oxfam; they are a set of logical machines,* each of which embodies a pure *Weltanschauung* having, in the eyes of the analyst and the actors, some relevance to Oxfam and to the task of managing it.

Because a main concern is information flow, each activity model can now be used to generate a model of the information flows needed by, and generated by that activity model. (Thus: the model components are activities captured in verbs; for each verb we ask: what information would be needed by anyone doing this activity? From what source? In what form? With what frequency? And we ask similar questions concerning the information generated by doing an activity.)

Back in the real world - but interested managers have themselves participated in the model building - we now compare the information flows generated by the models with actual information flows to the monthly management committee meeting. The models have not necessarily generated designs; they provide a coherent structure for a debate, one which is clear about what it is discussing and can examine previously implicit assumptions and generate new perceptions - a debate which provides a "self-organising system" with alternative agendas for change and an examination of standards by which judgements are made or have been made in the past, or could be made in the future.

The debate generates a need for new 'pure' system models: a system to publicise Oxfam work; a system to use modern information technology to service a committee. The new models are built; the debate continues; learning continues.

Eventually the debate creates the *readiness* to modify information flows in a particular way; the new problem situation becomes that generated by the decision to take this action. The cyclic learning process can begin again, and SSM has been used as an enabling system to catalyse this purposeful self-organisation in a social context.

Reflections on SSM and its Development

1. The nature of systems thinking is such that as soon as SSM has been developed and conceptualised as a (systems-based) learning system, an obvious consideration is the nature of the system wider than SSM itself, namely *the system to use SSM.*

I resist the temptation to model that human activity system, although doing so would be easy enough, not least because a model at that level of generality would undoubtedly be taken to the normative. SSM in fact leaves room for the user to use it in a way which fits his or her own mental modes. Empirical observation of uses of SSM (Atkinson, 1984) notes how the methodology, beyond its basic constitutive rules (Naughton; see Checkland 1981b) is moulded by the personality and interests of its user. Wishing not to diminish this user autonomy, I characterise the system-to-use-SSM only loosely - but that characterisation has been found to be of great practical value. Any use of the methodology may be conceptualised as the interaction of a would-be "problem solving" activity with some "problem content". This interaction is brought about by someone in the role "client". The organisation of the use of SSM is done by someone in the role "problem solver"; the problem solver *is free to make choices of who to be occupying the third role: "problem owner"*. (One person may, of course, occupy one, two or all three roles.) This freedom speculatively to assign the role "problem owner" has been found to be very important in opening up the thinking, and distances SSM from, say, RAND systems analysis, in which the real-world client is always assumed to be the problem owner. Speculating upon a wide set of possible problem owners provides a rich source of possibly relevant root definitions.

2. SSM emerges from the work briefly outlined above as a learning system available to those self-organising actors who take part in a study - in particular those who take part in the debate stage articulated by a comparison between system models and perceptions of the real world. The intention grew during the development of the methodology that it should not be the skill of a new variety of expert but part of the normal skills of anybody wishing to make a coherent approach to real-world problems.

3. The learning process set in motion by SSM has the character of an hermeneutic circle in which initial perceptions lead to new perceptions which modify those made initially. Because of that there is no good reason to agonise over the initial choice of "relevant systems" to be embodied in root definitions and conceptual models. The answer to the often-asked question: "How can I be sure my relevant systems are relevant?" is: "You cannot; make *some* choice and *learn* your way to what is truly relevant in this particular human situation." Applied in management situations, SSM serves to articulate a process of "evolutionary management" (Malik and Probst 1982).

4. The outcome of SSM is often perceived as a consensus of the actors who take part in Stages 5 and 6. This perception probably derives from the unfortunate prevalence of the functionalist view of organisations which sees them as equilibrium-maintaining systems for achieving instrumental action. Organisations in SSM are seen as complex processes rather than entities, and the outcome of SSM is best perceived as *an accommodation* which enables purposeful action to be taken. In the debate stage, differences of *Weltanschauungen* among the actors will be revealed; they may change, but a single unitary view does *not* characterise human perceptions and processes. An "accommodation" does not eliminate, or seek to eliminate, conflict, neither conflict of perception nor conflict of interest; that is its virtue.

5. It should be noted that the methodology *as such* cannot dictate who should take part in the definition of changes in Stages 5 and 6. Neither can it dictate the scope of the root definitions used - the degree to which they are radical or reactionary. In principle, given the nature of SSM as an explorer of perceptions among concerned actors, participation in the debate should be as extensive as possible in order to ensure the consideration of a wide range of changes when finding the final accommodation. Similarly, if only narrowly conceived conservative root definitions are formulated then the comparison stage will resem-

ble a comparison of & which & ; no useful insights will emerge. These issues
are important, but do not lie *within* the methodology. In that respect, the methodology as such is as neutral as a knife. Knives can be used to cut up vege-
tables for the cooking pot or to cut up one's neighbour; the knife is neutral
where the use to which it is put is not. Thus it is possible in principle for
SSM to be used in an emancipatory or a reactionary way, and it is impossible to
establish in principle that it will always be used in one way or the other. Its
general nature as an exploration of conflicting perceptions, however, is consonant with an effort to widen participation in the debate stages as much as possible and to examine both radical and reactionary root definitions.

6. Considering the intellectual structure of SSM as it emerged from "action research" (of which Susman and Evered, 1978, provide a useful modern discussion)
it is not without interest that it is permeated by complementarities. They include the following:

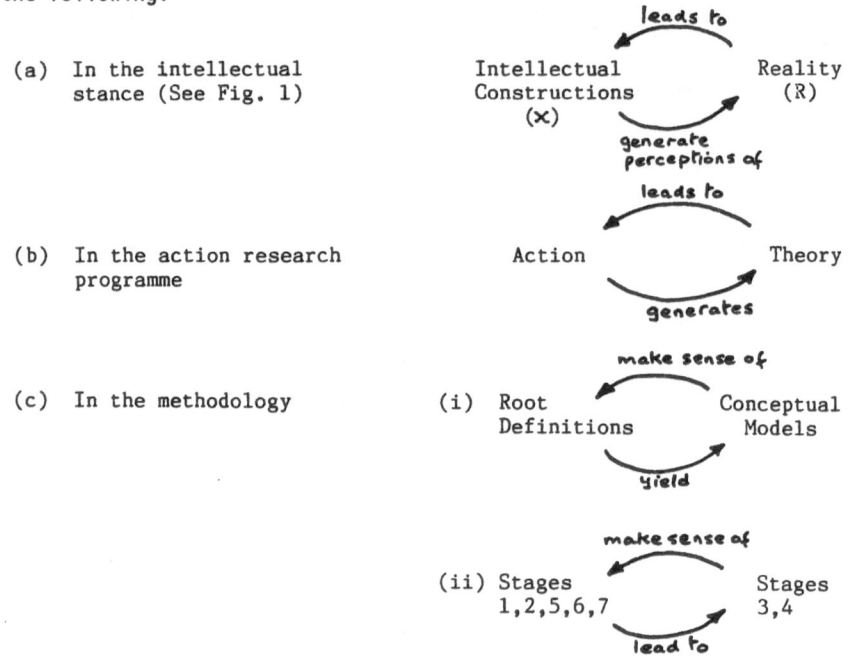

(a) In the intellectual stance (See Fig. 1)

Intellectual Constructions (x) — leads to → Reality (R) — generate perceptions of →

(b) In the action research programme

Action — leads to → Theory — generates →

(c) In the methodology

(i) Root Definitions — make sense of → Conceptual Models — yield →

(ii) Stages 1,2,5,6,7 — make sense of → Stages 3,4 — lead to →

This recurrence of groundlessness as an underlying organising principle alerts
us to similarities between the intellectual domains of this work and modern
thinking in biology (Varela, 1979; Maturana and Varela, 1980).

7. The notion of complementarity reappears in discussion of the model of social
reality implicit in SSM. (The issue is discussed at length in Checkland,
1981b). The image of the nature of social reality compatible with the experiences of using and developing SSM is one which maps interpretive sociology
(the Weberian tradition) and the phenomenological tradition in philosophy.
The model of social reality in SSM is this: that what in everyday language
we call "social systems" are the ever-changing outcome of a social process,
language-mediated, in which human beings continually negotiate and re-nego-
tiate with others their perceptions and interpretations of the world. The
interpretations generated themselves serve to constitute that which is interpreted, in a process (illustrated in Figure 5) akin to that described by Hejl
(1983).

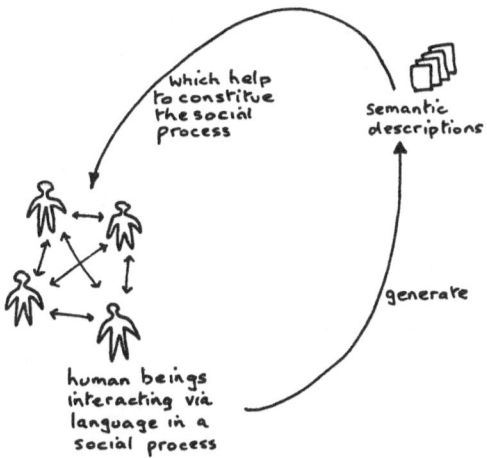

which help to constitue the social process

Semantic descriptions

generate

human beings interacting via language in a social process

8. The considerations in 6 and 7 above suggest that SSM could be used as a means of examining, for any given "social system", the drift through time of the descriptions which at a given time serve to constitute the social system itself. Thus might be addressed empirically the question as to whether the concept of an autopoietic system maps what we refer to in everyday language as "social systems".

9. The development of SSM has emphasised the fundamental difference between it and the tradition of "hard" systems thinking from which it has deviated. This can be expressed in terms of Figure 1 as follows (Checkland, 1981a, 1983):

- the 'hard' systems tradition assumes $\begin{cases} R \text{ is systemic} \\ M \text{ can be systematic} \end{cases}$

- the 'soft' systems tradition assumes $\begin{cases} R \text{ is problematical} \\ M \text{ can be systemic .} \end{cases}$

There is here a fundamental change of paradigm in which *systemicity is shifted from reality R to the process of inquiry into reality, M.* Both paradigms allow M to make use of x's which happen to be system models, but the nature of those models is different in the two traditions. In the "hard" tradition they are would-be models "*of*" parts of R; in the "soft" tradition they are models which it is hoped are "*relevant to*" discussion of R.

The development of the soft methodology has also drawn attention to the beguiling ease with which, when an x in Figure 1 appears efficiently to map part of R, we casually begin to talk (in everyday language) of x *being* that part of R. The everyday use of the word "system" in this way has made it almost worthless for use in professional discourse.

Finally this beguilement reminds us of the different kinds of error which can hinder discussion of matters of the kind raised here. Clearly, there will be errors of three kinds: the error of taking a description of an entity (x) to *be* that entity; the error of confusing a discussion about the rival merits of different descriptions (x1 versus x2) with discussion of the extent to which a given description (x) maps part of R; and finally the not uncommon error of mixing errors of the first two kinds.

If the special sophistication of those x's which are system models helps to indicate the prevalence in intellectual discourse of errors of the kinds described, then systems thinking will have made a contribution quite independent of its ability to provide a particular kind of x and a particular articulation of M.

References

Atkinson, C.J. (1984),   Ph.D. dissertation, University of Lancaster, in progress

Checkland, P.B. (1979),   Techniques in soft systems practice Part 2, J. Appl. Sys. Anal., 6, 41 - 49

Checkland, P.B. (1970),   Systems and Science, industry and innovation, J. Sys. Eng., 1, (2), 3 - 17

Checkland, P.B. (1981a),   Rethinking a systems approach, J. Appl. Sys. Anal., 8, 3 - 14

Checkland, P.B. (1981b),   Systems Thinking, Systems Practice, Chichester, Wiley

Checkland, P.B. (1983),   OR and the systems movement: mappings and conflicts, J. Opl. Res. Soc., 34 (8), 661 - 675

Chestnut, H. (1965),   Systems Engineering Tools, New York: Wiley

Chestnut, H. (1967),   Systems Engineering Methods, New York: Wiley

Hall, A.D. (1962),   A Methodology for Systems Engineering, Princeton, N.J.: Van Nostrand

Hejl, P. (1983),   Towards a theory of social systems: self-organization and self-maintencance, self-reference and syn-reference, Paper to the St. Galler Forschungsgespräche; written version to be included in this volume

Jenkins, G.M. (1967),   Systems and their optimisation, Inaugural lecture, University of Lancaster

Malik, F./Probst, G.J.B. (1982), Evolutionary Management, Cybernetics and Systems, 13, 153 - 174

Maturana, H.R./Varela, F.J. (1980), Autopoeisis and Cognition, Dortrecht: D. Reidel

Newell, A./Simon, H.A. (1972), Human problem solving, Englewood Cliffs: Prentice Hall

Smyth, D.S./Checkland, P.B. (1976), Using a systems approach: the structure of root definitions, J. Appl. Sys. Anal., 5 (1), 75 - 83

Susman, G./Evered, R.D. (1978), An assessment of the scientific merits of action research, Admin. Sci. Quarterly, 23, 582 - 683

Varela, F.J. (1979),   Principles of Biological Autonomy, New York: Elseview - North Holland

Vickers, G. (1965),   The Art of Judgement, London: Chapman and Hall

Evolutionary Management *

F. Malik

Institut für Betriebswirtschaft, Hochschule für Wirtschafts- und
Sozialwissenschaften, Dufourstrasse 48
CH-9000 St. Gallen, Switzerland

G. Probst

Institut für Betriebswirtschaft, Hochschule für Wirtschafts- und
Sozialwissenschaften, Dufourstrasse 48
CH-9000 St. Gallen, Switzerland

The only things that evolve by themselves in an organization are disorder, friction, and malperformance. (Peter Drucker, 1)

... the only possiblity of transcending the capacity of individual minds is to rely on those super-personal "self-organizing" forces which create spontaneous orders. (Friedrich von Hayek, 2)

1. Introduction and Basic Principles

There are no reasonable grounds for doubting that all forms of life which exist today and which are very successfully adapted to their environment constitute the result of an evolutionary process stretching over millions of years. Order of all forms of life and of their complex interaction patterns has evolved during and as a result of the evolutionary process, while at the same time, producing and determining the direction of that process, a direction that we can recognize with hind sight.

Even if the principles by which the evolutionary process works have not yet been researched and explained in every detail, the fact that natural organic systems evolve has nonetheless found almost universal acceptance.

The problem of socio-cultural evolution, on the other hand, is much more controversial. Notably the existence of social organizations which try to achieve goals by rational means is seen only in the context of rational human action on a basis of intentional planning and design.

Thus while natural systems are regarded as orders that have come into existence, or are spontaneous, social organizations are seen as made or constructed orders. This view is especially prevalent with relation to firms, which are regarded as comparatively efficient and purposeful organizations and whose design and control therefore seem to be based exclusively on rational, goal-oriented activities.

In this paper, we should however like to put forward the view that firms, like other social institutions and organizations, are self-changing, self-evolving and self-organizing systems, which are much less controllable, i.e., subjected to or open to the directing and designing influence of their managing bodies, than is generally accepted. Somewhat more precisely one could say that the controlling, designing or managing of a firm, as seen from an evolutionary point of view, displays other characteristics than those which it does if one starts from the basis that a firm is exclusively a rational system which was consciously and intentionally planned in a given way.

* Reprinted from: Cybernetics and Systems, An International Journal,
  13:153-174, Hemisphere Publishing Corporation, New York, 1982

Naturally the adoption of an evolutionary management concept does not mean that one must give up everything of the current state of knowledge about management and management methods. However, many things appear in a new light, and receive new interpretations; also, when seen in such a context, a whole series of problems become solvable, which, in the context of conventional ideas about management, do not appear to be solvable.

The concept of evolutionary management is not simply a new fashionable word. Of course, not everything is new. Many aspects of such a view will seem quite familiar to the experienced practitioner and possibly cause him to ask whether, for the observation of these things, a science is necessary, since it is anyway clear that reality is thus. This we would consider to be a strength of the evolutionary approach.

In business administration studies also, there is of course a whole series of approaches which point in an evolutionary direction, roughly the concepts and methods which have grown up in the context of the organizational development movement. Thus nothing would be easier than to set aside the evolutionary management concept with the remark that it is   old hat. However, the concept of evolutionary management in many respects transcends intellectually related approaches in business administration studies. It is based on a theory of social systems which, in its implications, reaches far beyond the limits of the firm, and which is also not completly new, but nonetheless distinguishes itself clearly from the dominant opinions in most economic and social sciences and has only recently received a systematic treatment in a few still relatively unknown works. (3)

Further bases of the evolutionary management concept (which we have space only to indicate here) are for example works from the areas of biology (4), psychology (5), comparative behavior research (6), sociobiology (7), the evolutionary theory of cognition (8), and an area that one can perhaps call ecological anthropology (9). Finally one should mention the real foundations in evolutionary theory, which already constitute an imposing body of human thought and understanding (10). A certain direction in cybernetics, which represents a direct continuation of the work of Norbert Wiener, but which has only very little to do with the regulation techniques or regulation theory perpetrated under this heading in also of considerable relevance. (11)

Already available results from the areas mentioned give ground to suppose that the evolutionary approach will lead to significant revisions of many opinions which have up till now been considered correct. Even if we limited ourselves just to examining approaches, we should still have no space here to go into a presentation of these results, so we should just like to stress that notably the role of the human mind or of human reason in connection with questions of human perception, cognition, judgement and design in all areas of life undergoes radical changes if viewed from an evolutionary point of view. It is a point of view which encourages modesty, caution and an awareness of the limits of the possible (12). This is true also and particularly for management as the most important social designing function.

The connection between management and evolution, or between management theory and evolutionary theory, is not at all obvious, and the forging of a connection between these areas may seem somewhat far-fetched to many people. What can the management of firms have to do with the process stretching over thousands of millions of years by which life in all its forms has become what it is today? What, in the knowledge about one of these subjects, could be of use in the other? Are these, things not so remote from one another that one could, at best, speak of insipid superficial and therefore in the last analysis fairly useless metaphors and analogies?

The answer to these questions depends on the point of view, and indeed mainly on the temporal historical and the political ordering points of view; finally this question is also connected with our conception of the nature of the firm. If one realizes that, in its subjectmatter, if not in its name, management is as old as the necessity to use the coordinated work of several people in order to carry out tasks which exceed the strength of a single individual, then it becomes clear that management in its many historical variances - from the forms of coordination used in the pre-historic hunt up to the landing of man on the moon - has in fact played a role in the context of socio-cultural evolution. Of course management, in this wide, all-embracing view, has neither determined nor controlled nor guided the process of socio-cultural development, as certain versions of history would suggest.

However, management has doubtless always been an important source of impulses, one of the many factors that have determined a direction taken by events, often without the carriers of these functions being aware of it and often quite simply because certain facts were created which made further development in certain directions impossible, but other developments easier or quite simply possible. A deeper understanding of the historical role of management than is usually to be found in today's managers would be desirable, not only for reasons of principle, but also because such knowledge is an absolute necessity for the strategic re-orientation of many firms, which is so much in the foreground today. Only thinking in larger time spans and on a wider horizon makes it possible to understand the present day situation and awakens the awareness not only of rapid and radical changes but above all of the changes which take place so skillfully that one grows used to them before one notices them, changes which have often been assigned to social and economic evolution. Up to now, it has been evident that every historical phase was only a transitory condition however permanent, enduring, desirable and worthy of defense it may have seemed to the people who experienced it. There is no reason to suppose that contemporary structures will be more enduring than their predecessors.

The second reason for the relevance of an evolutionary view point is directly connected to the first. Precisely because we must live with constant structural change and because in this firms or management play a decisive role, it seems important that management should also be aware of its current and future effects in the context of an ever farther reaching socio-cultural evolutionary process which is leading in an unknown direction and whose result is unknown. This evolution concerns above all the framework of economic activity, and it should not be overlooked that the naivity of many managers with regard to questions of organizational policy means that they are insufficiently equipped with a basic understanding of their business activity.

A further aspect, which is at least as important as the preceeding two, is that evolutionary theory in its contemporary form is not only an attempt to describe the actual steps in the process of evolution and not at all - as is often erroneously believed - the establishing of the laws of evolution, the knowledge of which should allow us possibly to foresee the future course of evolution. Evolutionary theory is above all the research of the mechanisms of evolution, of those operating principles which have produced the many forms of life which constantly astound us by their highly appropriate adaptation to their environment. It looks as if these operating principles might also be of considerable importance for the development of firms, and especially for their ability to adapt to unforeseeable changes.

The property of adaptability has not been of equal importance in all periods of history. In times of stable economic conditions, other factors may be stressed, and this may lead to a onesided specialization in the service of economic efficiency, which may turn out to be a threat to the existence of the firm, as soon as the conditions which have led to this development, for example a period of enduring

economic stability, cease to apply. Only then does it become clear that adaptability is the most important characteristic of a firm. The bankruptcies of recent
years illustrate this in a disconcerting way.

2. On the Relationship between System-oriented and Evolutionary Management

As has already been stressed, the aim of evolutionary management does not consist
in the developing of novelty for novelty's sake. On the contrary, this would review a lack of evolutionary thinking, since evolution is always based on building
on what is at hand, the keeping of the well tried, and on attempting further innovation from the starting point of the current state of affairs. If evolution had
begun over and over again, it would not today have surpassed the stage of single
cell organisms.

Evolutionary management proceeds completely organically from system-oriented management and thus represents a logical and handy further development of the latter.
As soon as one gives up a one-sidedly economistic point of view and attempts to
understand a firm as a system, the question quite naturally arises as to what the
orderedness of the system, in other words its systemic nature, is based on, how it
has arisen, how it can be altered and further developed.

Thus, an understanding of the firm as a system is an essential precondition to
being able to perceive its evolutionary character, as the first attempt to understand the actual relationships of cause and effect in a system shows that the complex organizational patterns to be found in every real firm can never be based entirely on conscious planning, organization, and management. They are in fact based
to a considerable extent on autonomous operating principles of the dynamics of the
system, which, if we understand them, we can use, but which we may also completely
strangle or destroy through ignorant interference with the system.

Evolutionary management presupposes a system-oriented viewpoint, for evolution can
be recognized and understood only with reference to the relevant connections within
a system; on the other hand, without an evolutionary viewpoint, the best that can
be imagined is a systematic but not a systemic conception of management. Complex
systems are so inseparably connected with ideas about change, adaptation, and development that one must remain incomplete without the other.

In this sense the evolutionary management concept is a consistent further development of the system-oriented approach in business studies, as it has grown up at the
Hochschule St. Gallen under the intellectual leadership of H. Ulrich. It has subse-
quently been further developed in numerous publications, partly theoretical, partly practice-oriented. (13) Parts of this evolutionary development have been subjected to a certain amount of practical trials, although one cannot yet speak of
empirical tests. (14)

3. The Firm as an evolving, self-organizing System

The evolutionary further development of a system-oriented management theory proceeds
from the insight that both the firm and its environment are subject to similar development processes and operating principles to those observable in natural evolution. At the center of this is the view that a firm together with its context is a
self-organizing system which can be organized and guided only to a limited extent
through conscious, planned intervention. This view is in contradiction with the
vast majority of dominant views in the worlds of science and practice. However,
this is not a sufficient reason to reject it lock, stock and barrel.

In what sense can one speak with sufficient precision of evolution and self-organization? Our approach rests basically on the following events: First, on a comparison of the ways in which alternative forms of coordination of human activity function; second, on the fact that human activity is governed by goals, but above all by rules of behavior; third, on the fact that activity in the social context often produces unintended and unforeseen secondary effects; and fourth, on the discovery by evolution-oriented social theory that there are exceedingly appropriate social organizations which are neither completely natural nor completely artificial, but belong to a third, largely ignored category, namely those systems which are indeed the result of human action but not of human intention, goals or planning.

Strictly speaking, these four basic elements are not independent of one another, but are rather different aspects of the same state of affairs.

3.1 Basic Forms of Coordination

We shall proceed from the contention that there are basically two different methods by which human action can be coordinated toward a common goal. The first variant consists fundamentally in coordination through command and instruction in the context of a hierarchical command structure; the second variant is the method of self-coordination in the context of a polycentric system (15) through reciprocal, anticipatory adaptation and modification of behavior by the participating persons or groups, or, expressed more generally, by the elements of the system.

The following example should make the distinction clear: the crew of a small ship, who basically act according to the captain's commands, represent a good example of a hierarchical command structure. Here the modes of behavior of the individual crew members, their relationships to one another, as well as their joint work are mainly determined, and in a fairly detailed way, by the commands and instructions of a central authority. On the other hand, the two teams in a football game are to a large extent self-coordinating and self-organizing systems. On the basis of the intention in both teams to win the game and of a possible basic strategy agreed in advance, the individual players behave to a certain extent as autonomous, decentralized decision makers according to the prevailing state of the game at a given moment. Their behavior is guided by the relative position of all the players and of the ball as well as by the commonly known rules of the game of football. In such a situation, commands or instructions by the team captain are hardly necessary, and one could even say that a team which depended on commands during a game would have only relatively small chances of winning.

The observation of these two forms of organization and of the coordination mechanisms connected with them is of course not new and is above all familiar to the economists from discussions about organizational policy.

It is noteworthy however that by far the majority of business studies and management literature concerns itself almost exclusively with the first form, that is to say with the many variations of command hierarchies and the appropriate methods for them. Forms of self-organization are treated only in special contexts, namely small group research, the study of informal organization and in the area of organizational development, and it must be pointed out that the explanations provided are often misleading for anyone who wants to understand the self-organization of the firm as a whole, because they refer to small groups. Small groups are of course important for self-organization, but their laws and rules cannot be transferred lock, stock, and barrel to larger systems. Although, in the context of self-coordination and self-organization questions relating to small groups, the really decisive advantage of self-organization is not thrown into relief, but rather problems of performance motivation and work satisfaction as well as humanization and the loosening of dominance forms.

The really decisive difference between a *system* based on the pattern of a command hierarchy and one based on self-organization and polycentric systems consists in the fact that a self-organizing system displays considerably greater adaptability than the command hierarchy. A polycentric system is able to process much more information and to perform mutual adjustment of a larger number of relations than the other type of system. This can also be easily proved quantitatively. (16)

If we are dealing with relatively simple relationships then the difference between the two systems may not be especially important. However, it is of existential importance as soon as the complexity of the task to be accomplished increases. If a firm is to survive and be efficient in a complex environment which is constantly changing in unforeseeable ways, it is necessary constantly to adjust and adapt such a large number of factors that this can be carried out only by polycentric, self-organizing system types. Systems which are organized and managed according to the model of the ship's crew become practically unmanageable even in circumstances of relatively little complexity, whereas systems based on the self-coordination model display a considerably greater abil.ty to overcome complexity and are thus more adaptable and show a greater ability to survive.

It would thus be fallacious to imagine that the two system types are genuine alternatives distinguished only by certain advantages and disadvantages, and that, in any given case, a choice can be made between them according to their relative suitability. Except for very simple cases, a system based on command hierarchy cannot solve the adaptation problems with which it is confronted in complex situations. It must be mentioned that if the information processing problem in a command hierarchy could be solved, then conceivably, because of various characteristics that it possesses, the latter system might produce better results and be more efficient and productive than a self-organizing system, which naturally, has a series of disadvantages, However, the problem is not a question of advantages and disadvantages, but one of the actually possible or impossible.

3.2 Order as a Result of Rules of Behavior

For a proper comprehension of the self-organization of a firm, however, one needs a further component (17), which we have so far only mentioned briefly in connection with the football game. In this context, we said that the behavior of the individual players is guided, among other things, by rules. The behavior of elements in a system is in fact not arbitrary, but is governed by rules which can be thoroughly understood by means of the analogy to rules of play.

We have no space to go into an exhaustive discussion of the existence, origins, and alteration of rules of behavior. We should, though, like to stress the following facts, which are basically results of socio-cultural evolutionary theory. Rules need not be intentionally made by anybody; rather they arise in the process of evolution as a result of the interaction of individuals with one another and with their environment, through a process analogous to mutation and selection. They are often not known to the persons concerned, but their functioning can be observed. Behavior rules in this sense are not arbitrary norms, but rather represent man's most important form of adaptation to an environment whose details he can never know sufficiently well to be able to guide his behavior according to the principles of cause and effect and to be able to behave rationally, in this causal sense.

Behavior rules tell man less what he should do than what he should not do, and thus limit areas where activity is safe or where its consequences and risks are at least able to be judged, from those areas about which too little factual knowledge is available and in which human activity therefore entails unknown risks and consequences. Behavior rules understood thus produce patterns of behavior, in other words an order of action. This order is, however, the most important pre-condition to being able to act intelligently, i.e., to being able to orient one-self on the basis of sufficiently stable data.

Behavior rules of play are in their way fixed, but are subject to evolutionary change, in the sense that, firstly, new adaptation requirements can lead to the creation of new rules, and that, secondly, groups which have unsuitable behavior rules will not survive. Thus socio-cultural evolution does not, as social darwi-nism asserts, work through the individual and his innate characteristics but selection refers to order of action (institutions) and culturally transmitted abi-lities and habits.

The rules which in fact apply, and which have been completed and modified by con-sciously created or imposed rules in relatively late phases of socio-cultural evo-lution,are one of the most important mechanisms of actual monitoring and regula-tion of a system. If these rules did not predate the attempt by any form of mana-gement to intervene, then the task of managing a system would probably be impos-sible. It is therefore also false to assume that there are social systems in the sense of societies precisely because men have started to shape their life together to a certain extent rationally with the help of consciously imposed ruTes, i.e., that there are rules because there is a society; rather, the exact opposite is the case: there are societies because rules were actually applied long before any con-scious norm was introduced.

3.3 Results of Human Action, but not of Human Design

The system of actions, events, relationships, regulations, etc., that is known as a "firm" is not entirely the result of conscious intention or logical planning. Although, possibly, nothing happens in a firm without human activity and although to that extent the situation concerned may be exclusively the result of human action, it is however usually not the result of human intention or design. Redu-cing it to a simple formula, one can say that the result of human action corresponds at best partially to the intentions of goals which determined the activity, and that quite often something results which was desired neither by any single indi-vidual nor by the whole group. (18)

This observation is completely reconcilable with the fact that every single one of the persons involved may possibly behave intentionally, logically, and in a goal-oriented way, and that the activity of all concerned may also be led by higher or more general common goals and purposes. Even in the best possible imaginable cases the above-mentioned effect may be observable. The examples of this effect are nu-merous. We shall limit ourselves here to the example of negotiations and discus-sions. Although every individual participant may enter negotiations or a discus-sion with clear goals and intentions, the actual result of negotiations will usual-ly be different from the goal that any single participant entered the negotiations with. The result of the negotiations is certainly the result of the behavior of all participants, but may be very far removed from their original intentions. Even the cause followed by the negotiations, the arguments, opinions, and suggestions made by any individual can - as every experienced negotiator knows - only be fore-seen within very narrow limits. Negotiation is thus a typical self-organizing pro-cess, whose decisive characteristic is that no participant has the negotiations as a whole under his control. In very many cases, it would probably be more appro-priate to say that the negotiations have the participants under their control, or that there is mutual control, where no party can be said to have clear or deter-mined dominance.

Putting it very generally, one can say that every situation in a firm which corresponds even only to a certain extent to reality is determined by so many influences that negotiators can never fully foresee the effects of negotiations, and that one is thus unable, when planning negotiations, to take account of all their possible effects. Actions in a social context will therefore always have unforeseen and unintended side effects, which will produce further actions on the part of all participants. In this way, there constantly arise new negotiating necessities, which nobody can plan in advance, but which nonetheless represent a reality.

If the effect of the unintended side effects and the related inherent dynamics of systems can already be observed in situations that are to a certain extent ideal, then naturally this applies all the more so as the proportion of actions that are not rationally determined (i.e., the proportion of spontaneous action and reaction) increases and the proportion of creative components in the negotiating also increases.

A clear statement that the results of human negotiations do not necessarily correspond to the intentions which led those negotiations will usually meet with the approval of many practicing managers, since it corresponds to their practical experience. This is not the least reason why one needs managers. One of their most important functions consists in constantly recognizing the situation as it now is, evaluating it, and, in the light of the most recent events, deciding and acting rationally.

It is perhaps not even exaggerated to say that this is the main task of managers. Certainly, a series of modern management methods such as planning, organization, etc., can contribute to reducing a little the area of uncertainty and of unintended side effects of every action,thus making it easier to see the situation as a whole and anticipate possible outcomes; however, these methods do not eliminate the need for action arising out of developments in the situation and foreseen by nobody. In the ability to act in the light of situations and yet guided by general behavior rules lies the real adaptability of a firm, and extensive use of certain methods, which in themselves may appear completely rational and logical, may destroy precisely this adaptability.

It would be a mistake to imagine that unforeseen side effects are always of a negative sort. Very often, they represent real opportunities, which are provided by so-called "chance" and which can be used. Every manager has had his experience and will secretly admit that many a success was not produced deliberately but resulted from a constellation of circumstances. A certain intuitive understanding of this effect, which is attributable to the natural dynamism of a system, is expressed, amongst other things, in the fact that managers sometimes consciously do not make impending decisions, because they feel the time or the matter itself is not yet right, and want to let events "have their head" for the moment; they do so in the hope that they will be able to carry out their intentions later, in changed circumstances, to which they have often contributed nothing at all or made only an indirect contribution. Doubtless, this procedure belongs, as a matter of course, to the repertoire òf experienced managers and politicians. Unfortunately, it is neither a part of the so-called theory of decision-making nor is it contained in management education.

Ways of behaving or "methods" of the kind we have just discussed are typical of relations with self-organizing, evolving systems. The example quoted here, of refraining from making a decision, is clearly not in all cases the expression of inability to make a decision, but rather an element of the indirect form of management which is often the only possibility of influencing self-organizing systems, and which uses their natural dynamics while leaving the latter intact.

Hence, waiting is not always "only" waiting, but often an active creating and ensuring of conditions which will provide freedom for the system to develop.

Although, unquestionably, this way of proceeding is used by experienced managers, one might argue that this is a rather unusual and atypical case. The view that one finally adopts depends entirely on the basic point of view that one has with regard to complex systems. What, in management, is still regarded as rather atypical, because people let themselves be led, to a certain extent, by the model of a "doer", is thoroughly familiar to ask from everyday life as a sensible method.

While we are aware that to guarantee the functioning of a machine we must know all its components and their joint workings down to the last detail, we know that as far as the rearing of a plant or an animal is concerned, this is not possible, and that the attempt as such would produce more damage than benefit. We limit ourselves to providing as best we can the conditions and requirements for its development, but otherwise we leave the system to its inner self-organization. Certainly, on such a basis, one can never foretell exactly the result which will in fact occur; we may, however, trust that the result will basically correspond to our expectations, even if it does not do so in every detail.

The same applies to the raising of children. Here too we have much less control over events than we like to imagine. We cannot command a child to love us or to respect us; we can only create conditions which favor the genesis of desirable reciprocal relationships.

Similarly, certain things which are partially necessary to existence cannot be commanded or made in a firm. Loyalty of fellow workers, identification with the firm, willingness to work and motivation can, in the last analysis, only be developed under favorable circumstances, but cannot be ordered. Therefore, in such and many similar cases management must refrain from "activistic" intervention in the inner functionings of the firm and limit itself to cultivating favorable conditions and supporting, as a catalyst, the natural development of certain desirable results and qualities. Precisely an abstention from pseudo-rational ordering of details gives us the possibility of attaining results which cannot be attained in any other way.

It is in no way a question, as is often insinuated, of a conceptless "getting by." Procedures of the kind we have discussed are, on the contrary, based on very clear concepts, which cannot be described in detail here. (19) In any case, the important thing is that precisely the self-organizing nature of the firm is an essential component of a concept of overall development and of development management.

Objections

Against the views outlined here certain objections can be made, which can be subsumed in four large groups.

1. The first group accepts the existence of the self-organizing natural dynamics of the firm, but pleads, precisely for that reason, for an increased use of methods of which it is supposed that they will, in the last analysis, enable people's behavior to be brought more and more under control. We shall summarize these arguments or objections under the name "automaton arguments" because, in one form or another, they are based on the idea that a firm can be managed in such an aware and planned fashion that it finally functions exactly like a machine. The thought expressed in this fashion of arguing can also be called the "constructor's or builder idea" or "technomorphic thinking".

This way of thinking especially characterizes business and management studies as scientific disciplines as well as vast parts of the management education that is based on them, and thus finds its way into management practice. However, other areas of the economic and social sciences are also heavily influenced by such ideas, roughly certain areas of national economics, of psychology and of administration sciences. Lastly, a similar tendency can also be discerned in legal sciences.

The technomorphic way of thinking has its roots, on the one hand, in the undisputable successes of some natural sciences and of the resulting technology and, on the other hand, in some deeply rooted ideas about the functioning of society, especially in the model of a closed tribal society, in which the main problem consists in the satisfaction of known needs of known people, and the relationships and circumstances are so clear that the problem can possibly be solved with rational, goal-oriented means. The strong emotional attractiveness often attached to such ideas perhaps results from the circumstance that mankind lived for a very long period in such societies and, even today, spends a considerable proportion of his life in such social systems, in small groups and above all in the family.

It is not possible, within the context of this paper, to attempt a complete refutation of what we have here called the "automaton argument". We must content ourselves with referring the reader to certain important recent works on the subject. (20) We should just like to stress that it is the constructor's way of thinking, in its many variance, that is the biggest hindrance for an adequate understanding of the self-organizing characteristics of a firm and consequently of the development of a theory of design and control of social systems which does justice to reality. The almost insuperable belief in the feasibility of all things in the undoubted efficiency and superiority of human reason with reference to the solving of organizational and management problems in the manageability of people and organizations makes it impossible for many people to see where the limits of the feasible lie and how limited are our possibilities for organizing things according to our ideas.

The situation is perhaps comparable with the time when men's main and (literally) visible experience raised no doubts about the supposed fact that the earth was a ' flat disk. Even today, our everyday experience confirms this idea ...

2. The second objection which is often raised with regard to the self-organizing nature of a firm consists basically of the idea that even a well thought out plan and an ever so carefully balanced and considered concept can nonetheless be frustrated by other people. In the context of this particular argument, the fact that the result of human action very often does not correspond to our intentions and goals is interpreted as a conscious and intentional engineering of mistakes and setbacks. Continuing a certain condition, we shall call this argument the "conspiracy argument" (21), or also the "scapegoat theory". It is often used in conjunction with the automaton argument to explain the failure, in individual cases, of so-called rational methods, and is probably the most common form of (pseudo-)explanation of social facts and events.

This way of arguing, which for most people has a tremendous degree of plausibility and persuasiveness, is based on the conviction that every event has its causes and that it must be the result of planned intentions. Consequently, if any event conflicts with our own interests, it is, after all, only obvious that we should look for those people who could have an interest in this event and who therefore might possibly have engineered it. In this way of looking at things, there is no room for the view that even useful events or states, which de facto serve the interests of certain groups, might be the *unintended* result of the combined interactions of the most carefully planned actions. There is an immediate search for "stream-pul-

lers" or for "scapegoats", and all observations are interpreted exclusively in
the context of such an approach.

Of course, it cannot be disputed that there are conspiring groups of interest, and
that there are intrigues for and against absolutely everything and also conscious,
energetic attempts to frustrate many plans both inside and outside a firm. But it
must be as freely admitted that many intrigues and conspiracies fail to reach their
intended objective. And even if they interfere considerably with our own intentions,
or even produce a complete change of these intentions, the outcome is rarely one
which was thought by any of the participating persons or groups. Every intrigue is
also situated in a social context and produces reactions of an unforeseeable number
and type, so that a person who intrigues not infrequently becomes the victim of
his own intrigues and conspiracies often escape from the control of the conspira-
tors. World history is full of examples of such happenings.

Thus the conspiracy argument is not a valid objection to the theory of self-organization as applied to firms, but simply prevents those who use this argument from
perceiving the realities of a situation. Because of its superficial persuasiveness,
this argumentation syndrom can, however, be used to great effect as a means of
self-defense, i.e., to justify one's own failures and to build up pictures of an
enemy.

3. A third objection, which also appears in many variants, consists of a denial
that the phenomenon as such exists, and that it is a kind of psychological self-
deception, which takes place partially without the person being aware of it, rather
in the same way as optical deception occurs. However, this "trick" is also used
consciously on such occasions, the persons using this trick not infrequently end
up by starting to believe their own assertions. We affirmed the simple fact, often
encountered in every day experience, that many people behave post facto as if what
always correspond to their well thought-out plans. This is a form of what psycho-
logists but above all psychiatrists know well as expost-rationalization and which
is excellently expressed in the fable of the fox and the grapes.

This reaction of harmonizing goals post facto with the actual results very effec-
tively prevents people who experienced it from perceiving the self-organizing character of social systems. Here, as much as it is a consciously used social mani-
pulation technique it is of course indisputable that it is initially very effective, but it usually has, in the long term, very destructive effects, because it
is well suited to undermining the foundation of trust which is essential for all
cooperative action. This behavior destroys the basic rules of play of social collaboration and is therefore rightly felt to be unfair.

4. The fourth group of objections consists in a dismissing of the observation that
the firm is a self-organizing unit as trivial or irrelevant. Such a view is based
on the argument that it is of course obvious that organizations consisting of
human beings cannot function one hundred percent perfectly and that one must simply live with the manifold frictions and breakdowns in the system. This attitude
comes, in a certain sense, the closest to reality and is also benign in as much as
it does not produce the distortions of perception that the other objections lead
to.

However, because of the trivialization frequently connected with this view, people
fail to realize that self-organization is a fundamental and important property of
social system; they thus give up the possibility of gaining and understanding of
the nature of a firm which could be qualified as really sound and appropriate to
reality and which is the best method for improving the structure and management of
such systems.

In fact, in the case of events and circumstances which deviate from the intentions of the persons involved, what one is looking at is neither the work of conspiratorial circles or intrigues nor simply a breakdown of the system. Rather, it is very often the outwardly recognizable sign of adaptational processes (arising from the dynamics of a system itself) to circumstances which are unknown to anybody in their totality.

5. What are the Lessons for Management?

5.1 More Insight

It is always easier to demand recipes than to strive to attain an insight. However, it is insight into the real functioning of firms which is really needed, and it is such insight which is also lacking in management training.

Drucker (22) expresses this in a fashion which can hardly be improved upon

> The bulk of the work today concerns itself with the sharpening of already existing tools for specific technical functions - such as quality control or inventory control, warehouse location or freight-car allocation, machine loading, maintenance scheduling, or order handling. And, in fact, a good deal of work is little more than a refinement of industrial engineering, cost accounting or procedures analysis and improvement of functional efforts - primarily those of the manufacturing function but also, to some extent, of marketing and of money management.

> But there is almost no work, no organized thought, no emphasis on managing an enterprise ...

Since then the situation has changed only slightly. A few new techniques have been introduced, the centers of gravity have moved slightly, but the question of the nature of a firm and of its actual working principles, from which, in the last analysis, a   sensible use of techniques and methods depends, still plays a far too subordinate role.

Thus, the realization is only slowly dawning that the highest goals of a firm cannot be profit and growth of turnover but that only an orientation towards the enduring ability of the firm to survive offers any chance of making correct decisions or rather avoiding wrong ones. In the stable years of the 50s and 60s, which were characterized by continuous economic extension, the superstition became widespread that growth of profits and turnover was the decisive goal, with the result that not only were all management techniques aligned on this idea, but also that a whole management "culture" has grown up around this idea.

This has induced what is possibly the most dangerous weakness of our economic system, and which, until now, has not been recognized: it is the fact that practically a whole generation of managers has had no opportunity to gather experience of how to manage in times of economic crises, and therefore, in many cases, can clearly be seen to be helpless in the face of changed circumstances.

Another question, which also can be sensibly answered only on the basis of deeper insight, namely the problem of how, in a firm, one creates such subtle but "essential goods" as loyalty, identification, confidence and interest, is hardly mentioned in management theory and management training. These things cannot be commanded, but must grow. We can never produce them by direct intervention, but can only hope to influence certain prerequisites for their genesis.

5.2 Better Methods and Solutions

However, an evolutionary management concept not only provides new insight. Doubt-less, better, evolutionarily tested methods can also be expected. One could think for example of the obvious analogy between an ecological niche which makes possible the survival of a species (23) and a market niche, which has the same effect for a firm. Another example would be the proven optimum usefulness of a trial and error strategy for the solving of complex problems. (24)

For certain areas of technology, it has already become a matter of cause [under the name "bionics", (25)] to use solutions evolved in nature as models for the so-lution of technical problems. On this basis, an exceedingly interesting structural model of the organization of a firm has meanwhile also been developed, which may represent a real breakthrough. (26)

5.3 Better Understanding

Thanks to an evolutionary approach, a better understanding of the ecological role which every firm per force has to play is made possible. By this is meant less the question of protection of the environment than the problem of the positioning of the whole firm in a complex network of socio-cultural, technological, economic and political factors and influences. Ecosystem research (27), which examines the genesis, structure and dynamics of such networks of effects, will possibly be more important in the future for business management than is political economy.

In other words, it is conceivable that biologists and behavioral researchers may, in the last analysis, be able to develop a more correct understanding of the forms of competition and collaboration than can political economists, for whom firms are, in the first analysis, economic, i.e., unfortunately, fictitious entities.

In the future, it will, however, be above all a question of understanding the eco-logy and evolution of ideas and value judgements. In ten years' time, an ecology of the mind (28) and the emotions will perhaps be one of the obvious subjects in a course of management training. After all, no one can doubt that it is ideas that have changed the world. The dissemination of knowledge and values, the laws accor-ding to which ideas, values and feelings evolve, are selected and transmitted, are vastly more important than knowledge of product life cycles, technological substi-tutions and the like, since ideas, values and feelings are at the root of all deve-lopment.

6. Learning to be - what we are (29)

We have attempted to set out the most important basis of an evolutionary conception of management. We have limited ourselves to elementary insights and have neglected special concepts from the area of biology or of evolution theory. This is justified by the fact that, in the present state of dissemination of knowledge and insights relating to evolutionary theory, there cannot initially be any question of trea-ting complicated cases and special phenomena, but rather those bases that are ab-solutely essential and are thus the prerequisite for even being able to perceive evolutionary and self-organizing behavior, the prerequisites for the awakening and the sharpening of our senses towards these things.

The abstention from using special concepts, which would have made many things ea-sier to write, but probably not to read or to understand, is not only required by the need to be comprehensible, but is also justified by the fact that the tradition of evolutionary thought is much older in the social and economic sciences than in biology. One may suppose that without the great theoreticians of evolution in the social sciences in the 17th century (30), Darwin's theory would simply not have been imaginable. The continental European variant of constructive rationalism absorbed thoroughly the then established truth that we must today, with shame, relearn from biologists some of the most basic elements of knowledge relating to the social sciences, and this in a language which is not particularly appropriate to social reality.

However that may all be, the main lesson of evolutionary theory is that we are part of a permanent process of development whose future pattern we cannot foresee, but to whose direction we can make a small contribution, even if it be only through the fact of our existence. As managers, we have from time to time a chance to exercise a more intensive influence on the direction of development, provided that we learn to be what we really are: not doers and commanders, but catalysts and cultivators of a self-organizing system in an evolving context.

Notes and References

(1)   This does not mean that P. Drucker only has an anti-evolutionary viewpoint, see also Drucker, F., Management, London 1973, p. 637 ff.

(2)   Hayek, F.A., Law, Legislation and Liberty, Vol. I, Rules and Order, London 1973.

(3)   See Hayek, F.A., Law, Legislation and Liberty, Vol. I-III, London 1973-1979, Studies in Philosophy, Politics and Economics, Chicago 1967, New Studies in Philosophy, Politics, Economics and the History of Ideas, London 1978. Nozik, R., Anarchy, State and Utopia, Oxford 1974. Oakshott, M., On Human Conduct, London 1975. Hayek, F.A., The Sensory Order, London 1976.

(4)   See Riedl, R., A Systems-analytical Approach to Macro-evolutionary Phenomena, in: The Quarterly Review of Biology, 52, p. 351-370; Varela, F., Maturana, H.R., Uribe, R., Autopoiesis: The Organization of Living Systems, Its Characterization and a Model, in: Biosystems, 5, p. 187-196; Varela, F., Principles of Biological Autonomy, New York 1979, Maturana, H., Varela, F., Autopoiesis and Cognition, Dordrecht, Boston 1978; Zeleny, M., Autopoieses: A Theory of the Living organization, New York 1980, Waddington, C.H., Tools for Thought, New York 1977, The Strategy of Genes, New York 1957.

(5)   See all the numerous publications of J. Piaget but also Weick, K.E., The Social Psychology of Organizing, Reading Mass. 1979[2], Enactment Processes in Organizations, in: Staw, B.M., Salancik, G.R., (Ed.), New Directions in Organizational Behaviour, Chicago 1977.

(6)   See Lorenz, K., Behind the mirror, London 1977, On Aggression, London 1966.

(7)   See Wilson, E.O., Sociobiology: The New Syntheses, Cambridge, Mass. 1975; Jantsch, E., Waddington, C.H., Evolution and Consciousness, Reading, Mass. 1976, Maruyama, M., Toward Cultural Symbiosis, in: Jantsch, E., Waddington, C.H., Evolution and Consciousness, Reading, Mass. 1976.

(8)   Eccles, J.C., The Human Psyche, Berlin 1980, Facing Reality, Berlin 1970, The Human Mystery, Berlin 1978; Riedl, R., The Biology of Knowledge, Chichester 1984, Popper, K., Objective Knowledge - An Evolutionary Approach, Oxford 1972, The Logic of Scientific Discovery, London 1959; Popper, K., Eccles, J.C., The Self and Its Brain, Berlin 1977; Von Foerster, H., On

Constructing A Reality, in: Preiser, W. (Hrsg.), Environmental Design Research II, Stroudsbourg 1973; Pask, G., An Approach to Cybernetics, London 1972, Conversion-Theory: Applications in Education and Epistemology, Amsterdam 1976; MacKay, D.M., Brains, Machines and Persons, London 1980, Information, Mechanism, and Meaning, London 1969.

(9)   See Bateson, G., Steps on an Ecology of Minds, New York 1972, Mind and Nature, New York 1979.

(10) See Eigen, M., Self-Organizing of Matter and the Evolution of Biological Macro-molecules, in: Naturwissenschaften, 58, p. 465-523; Eigen, M., Schuster, P., The Hypercycle: A Principle of Natural Self-Organization, in: Natur-wissenschaften 64, 1977, p. 541-565, 65, 1978, p. 7-41, 341-369, and also published in a book, Berlin 1979; Dobzhansky, T., Ayala, F., Stebbins, G., Valentine, J., Evolution, San Francisco 1977; Darlington, C.D., The Evolu-tion of Man and Society, London 1969; Jantsch, E., Design for Evolution, New York 1975, The Self-Organization of the Universe, New York 1980.

(11) Beer, S., Decision and Control, London 1966; Brain of the Firm, London 1972, Platform for Change, London 1975, The Heart of Enterprise, London 1979.

(12) See Riedl, R., The Biology of Knowledge, Chichester 1984

(13) Nearly all the publications about the St. Gall Systems Approach are in Ger-man. See publications of H. Ulrich, W. Krieg, P. Gomez, F. Malik, G. Probst; and Note 19 ; See also Gomez, P., Malik, F., Oeller, K.H., Organic Problem Solving in Management: A System-Methodology, in: Proceedings of the Third European Meeting on Cybernetics and System Research, Vienna 1976, Malik, F., Systems Ideas and Managerial Practice: A Swiss Experience, in: International Cybernetics Newsletter 1/78, Gomez, P., Top Down versus Bottom Up Organiza-tional Design: A Cybernetic Perspective, in: Journal of Enterprise Manage-ment, Vol. I, pp. 229-239, Organic Problem-Solving in Public Administration: A Systems-Methodology, in: Sutherland, J., Management Handbook for Public Administrators, New York 1978.

(14) See Malik, F., Management-Systems, in: Die Orientierung, Bern 1981 (in German).

(15) See Polanyi, M., The Logic of Liberty, London 1951.

(16) See Polanyi, M., The Logic of Liberty, London 1951, p. 114 ff.

(17) See Hayek, F.A., Law, Legislation and Liberty, Vol. I: Rules and Order, London 1973.

(18) See Hayek, F.A., Law, Legislation and Liberty, Vol. I: Rules and Order, London 1973.

(19) See Probst, G., Cybernetic Laws as a Basis for Rules of Design and Control in Management, Bern 1981 (in German), Gomez, P., Probst, G., Cybernetic Rules and Management-Principles, in: Ericson, R.F. (Ed.), Improving the Human Condition: Quality and Stability in Social Systems, Proceedings of Society for General Systems Research, Berlin 1979, Centralization versus Decentrali-zation in Business Organizations: Cybernetic Rules for Effective Management, in: Cybernetics and Systems, 11, 1980, p. 381-400.

(20) See Hayek, F.A., Law, Legislation and Liberty, Vol. I: Rules and Order, London 1973.

(21) See Popper, K., Objective Knowledge - An Evolutionary Approach, Oxford 1972.

(22) See Drucker, P., Technology, Management and Society, New York 1977, p. 1972.

(23) Wynne-Edwards, V.C., Animal Dispersion in Relation to Social Behaviour,
      New York 1962; Mayr, Ed., Animal Species and Evolution, Cambridge, Mass.1963.

(24) See Rechenberg, J., Evolutionsstrategie, Stuttgart 1973 (in German).

(25) See Von Foerster, H., Biological Ideas for the Engineer, in: New Scientist,
      15, 1962; Von Foerster, H., Bionics Principles, in: Willamne, R.A. (Ed.),
      Bionics Lecture Series XX, Vol. 1, NATO Advisory Group for Aerospace Research
      and Development, Paris 1965.

(26) See Beer, S., Brain of the Firm, London 1981[2]. See Beer, S., The Heart of
      Enterprise, London 1979.

(27) See Vester, F., Urban Systems in Crisis, Stuttgart 1976, Vester, F., Hesler,
      A. von, Sensitivity-Model, Stuttgart 1981.

(28) See Bateson, G., Steps to an Ecology of Mind, New York 1972, Mind and Nature,
      New York 1979.

(29) See Beer, S., Decision and Control, London 1966, p. 355 ff.

(30) We think of Adam Smith, Adam Ferguson, Bernard Mandeville, David Hume.

Systems Approach to Management: Hopes, Promises, Doubts –
A Lot of Questions and Some Afterthoughts

F. Malik

Institut für Betriebswirtschaft, Hochschule für Wirtschafts- und
Sozialwissenschaften, Dufourstrasse 48
CH-9000 St. Gallen, Switzerland

By choosing general systems theory and cybernetics as a basis for a new sort of
management theory many more problems arose than were solved. Let me give you a
very personal account of how I see our endeavours to understand and use systems
sciences and thereby of the state of our understanding. Expectations were very
high but the obstacles proved to be very difficult to overcome:

1. It was necessary to learn a completely new language and to acquire new ways
   of thinking for the adoption of which both management scientists and management practitioners were and are badly prepared.

2. The classical perspective of a one-dimensional economistic sort of micro-eco-
   nomics (Betriebswirtschaftslehre) and the uni-directional causal thinking of
   classical management theory dominated almost everything that was going on in
   these areas. It proved not only very difficult and almost impossible to identify the hidden assumptions and premises of these approaches but the main problem
   was that on the basis of such premises world pictures and world realities were
   constructed by the very conduct of managers which in turn was the best and
   irrefutable proof of the correctness of their ways of thinking.

3. At a closer look general systems theory itself appeared to be of only limited
   help due to its inherent problems which mainly resulted from its very diverse
   and irreconcilable approaches. The only unity in this allegedly unifying science seemed to be the very frequent use of the word "system".

However, a certain kind of cybernetics seemed to be particulary interesting and
useful for management. It developed around scientists who partly became almost
legends during their life time. I would like to mention only a very personal selection: Ross Ashby, Warren McCulloch, Stafford Beer, Gregory Bateson, Gordon Pask,
Friedrich von Hayek, Jean Piaget, and of course also our guest Heinz von Foerster
who not only is a cybernetician but also was some sort of gubernator of all these
cyberneticians, in so far as he succeeded in managing one of the most fascinating
research groups in this area, in which most of these scientists participated in
some way for some time. Irrespective of whether these men would accept being called cyberneticians or whether they themselves did use the term cybernetics in their
written works, their thinking seems to show a special kind of integrity and unity
not only with respect to each one's work but also between them:

1. There was explicit acknowledgement of extreme complexity of systems which was
   understood to result from the interaction of their parts as well as from their
   interaction as whole systems with other systems.

2. There was a clear and sustained effort to treat systems as integrated wholes
   and to avoid the fallacy of reductionism and analytic desintegration.

3. Nevertheless and certainly not in contradiction with anti-reductionism there
   was hope for a new science of simplification which by preserving the integrity
   of systems would allow a better understanding of their nature.

4. There were some very interesting theorems of insolubility of certain problems by application of certain methods, for example Bremerman's Limit.

5. Another unifying and common strand seemed to be a special understanding of the problem of control, and this was the real point of interest for both management scientists and management practitioners.

Almost all important management problems seem to be related to the control phenomenon as these cyberneticians understand it. Keeping all sorts of operations on course, be it production, marketing, financing,etc. but also all control processes of operations like planning, goal setting,etc. seems to be in essence control phenomena. Adaptation, learning, development, evolution appeared to be understandable as particular classes of control phenomena. Most recently one could even learn that all these things are closely related or perhaps even identical with a sufficiently broad understanding of the concept of computation, so that we seem to be entitled to say that management is the computation of computations ...

One of the most interesting and most important results of this kind of cybernetics is the insight that control cannot be separated from the system but is intrinsic to it, that it is implicit in the system's structure. From this follows that it is not the manager, nor all the managers taken together, who constitute the control mechanism of a system. So if we take seriously this sort of cybernetics as a basis of management, or, to formulate it the other way round, if a company or any other social institution is seriously taken to be a system, then this leads us unavoidably to the idea of self-control and of self-organization. If systems are not really but only apparently controlled and/or organized by management then systems control and organize themselves.

The idea of self-organization seemed to open a path way to a really new and very different understanding of management, of institutions, and of managing an institution. Whereas on the grounds of first-order cybernetics the manager could still uphold the delusion that it is he who installs feed-back control systems, establishes regulatory criteria, modifies transfer functions, sets goals, takes decisions, and uses all this as a new technology all on the way to the automated factory which would then have to be called cybernated factory since the manager would have organized the system to regulate itself, nothing of this kind can be upheld any longer if we really accept complex systems to be truly self-organizing.

At first look this was fascinating, and a liberation of one's mind from the straight jackets of an authoritarian theory of authoritarian management trying co-operatively to persuade people to its authoritarianism.

But at second look the hydra of self-organization exposed its power to proliferate mind racking problems. The way to the heaven of self-organizing systems and to better theories of management is plastered with the nightmares of ever more complicated theories of self-organization and instead of heaven one finds oneself very soon in the hell of paradox, self-reference, contradiction, and an almost solipsistic and monadic epistomology. Dangerous things, indeed, although it seems that the theory of autopoiesis might save us.

There were extremely complicated theories of neuronal networks and their immanent logic; there was a paper about self-organization co-authored by Heinz von Foerster which used a funny picture of a man with a bowler hat *on* his head and several other persons *in* his head to show how solipsism is to be avoided.

There was at about the same time a paper by one of the real pioneers of cybernetics (Ashby) who came to the conclusion that there is no real self-organization. The man who according to the references made the drawing of the man with the bowler hat, Gordon Pask, and whose earlier theories of self-organizing teaching/lear-

fig. 1

ning machines were still quite understandable, invented ever more complicated theories of conversational systems. And last but not least another one of the pioneers (Beer) developed a model of any viable system. A further stream of thought comes from the theory of biological evolution of ethology and of social and cultural evolution which sometimes appears to have little in common with those theories already mentioned, but are clearly theories of self-organization.

And above all for the social scientist there is the scientific work of Friedrich von Hayek with the theory of spontaneous, self-generating, and self-maintaining orders, the self-maintenance of which results from the behavior of millions of elements - individuals - acting according to rules which they do not necessarily know, following their own individual and local purposes, thereby giving rise to an order of relations which they do not know, in any case not in its entirety and whose only purpose, if we are allowed to use this word at all, is to enable the individuals to interact in an order-preserving way, the result of which we could also describe as survival of the order. In my opinion this theory is very similar to the most recent work in the field of self-organization, namely the theory of autopoiesis without, however, drawing the same far-reaching conclusions as this theory does.

In the light of the theory of autopoiesis we are facing some real dilemmas. Those of us who uphold a systems theory which is based on the idea that systems are open and complex, that they exhibit goal directed, purposeful, and meaningful behavior, that they are free to choose, and decide, that they make representations of their environments by gathering information and thereby learn something about its order and that they communicate by exchanging information by means of language have to cope with statements of the following kinds which I selected from the writings of Maturana and von Foerster:

- that recognition of order and complexity is a property of the observer and not the observed;

- that living systems are not open but informationally and operationally closed and strictly deterministic in the sense of state determinateness - things one could already have learnt from Ashby as early as 1956;

- that it does not make any sense to talk about purpose and meaning and goal directedness except in the language of the observer which has nothing to do with the observed system;

- that it is untenable to say that nervous systems are organs which code or store or gather information by means of a sensory apparatus from an environment about that environment and that we have to give up the idea that it computes an environment;

- that language does not transmit information;

- that nervous systems are not hierarchical systems and that the cortex is not a center of "higher functions" neither of mammals nor of men.

Almost everything which most of us take to be of value from the theory of self-organization for a theory of management seems to collapse in the light of these statements.

Those who are upholding the perspective of open, complex, goal-directed, purpose-ful, and meaningful systems as described above use according to the theory of auto-poiesis a description of the observer and all those properties are properties of this description and not of the systems thereby described.

For some time I thought that the distinction and use of the concepts of object language and metalanguage could serve as a solution. In a way it does. It helps us to recognize that whatever a manager or control system does is done from a meta-systemic level. This is important for them to understand since otherwise they get overwhelmed by the proliferating complexity of the system to be controlled and end up at Bremerman's Limit.

But the theory of autopoietic systems seems to imply that the concepts of object- and meta-language or object- and metasystem are not an entirely satisfying solu-tion. It rests on an artificial disconnection of elements which are in fact connec-ted through a network. Every would-be controller must be connected to the system he tries to control and therefore becomes part of it. This constitutes an additio-nal loop in a network which gets closure again. The network remains heterarchical and in what way the new loop changes its characteristics remains to be seen. It can not be calculated or predicted.

If we look at a system from this point of view it does not really have an environ-ment in the sense in which we generally use this term. It then is a system with operational closure.

What could this mean for the management of an enterprise? Where or what is the be-ginning and the end of such a system? Where are its boundaries? In principle this would entitle us to take any element as focus of an analysis as long as we do not break the closure; as long as we close in on the element itself as is shown in fig. 2.

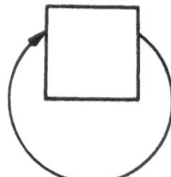

fig. 2

In the managerial context, however, it seems most appropriate to use the customer as the focal element. This means to understand the entire loop from the customer to the customer (fig. 3).

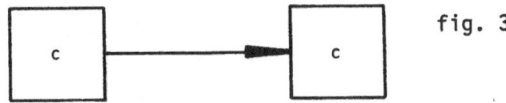

fig. 3

What lies between the two squares and is represented in a very over-simplified manner by an arrow is in fact the entire and very complex network that constitutes the enterprise together with everything else that influences the behavior of the customer (competitors, suppliers, etc.).

Both little boxes stand in fact for the same customer so we can depict it as in fig. 4.

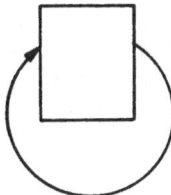

fig. 4

Useful as this modelling may be for a general understanding of the basic nature of such a system it is certainly not sufficient.

If we recognize, however, that whatever happens in an enterprise contributes in the last consequence to the closure of the system, we can develop a rather unusual way to analyse it and in consequence to establish or improve its intrinsic control and coherence. I will try to give a very abbreviated account of the major steps of such a method.

Step 1: Take representative groups of managers and workers of an enterprise. Let them depict the system of which they are directly responsible in the form of an input-output flow diagram in which their personal function is the centerpiece of the network. Demand that they adopt a particular viewpoint in doing this analysis, namely, that they answer the following questions:

1. What is my contribution to the enterprise?

2. What are my outputs and who are the receivers?

3. What do I receive back? (Note: No undirectional arrows are allowed in the diagram)

4. What are my inputs and who are the suppliers?

5. What do I give back?

In this way we will get as many closed systems diagrams as there are participants.

Step 2: Let every participant present his diagram graphically and verbally to all other participants simultaneously, so that each one of them can correct the functional lay-out of every diagram.

By this we get a set of diagrams which represents every individual perspective which is at the same time modified through all the other perspectives.

Step 3: Project all these diagrams on to each other so that identical elements take identical places in the network. Since all the diagrams are models of the system we will arrive at an entire systems model which has been made by the system itself of itself. By this method one gives the system only a medium, a language, through which it can express itself, but one does not predetermine what the system will take as relevant to its own nature.

However, since there will be also those elements involved which have direct contacts with the customer, it will be a model of the system from the customer to the customer by which the system defines itself. At the same time it is a model of how the system produces itself, by closing in on the customer which is its purpose of existence.

Everything we need to take into account is enclosed in the model; there is no outside world; the system is operating recursively on its own operation. The question which is left is whether it will have Eigen-values. If yes, it will survive; if not, it will vanish.

Cybernetic Principles for the Design, Control, and Development of Social Systems and Some Afterthoughts

G. Probst

Institut für Betriebswirtschaft, Hochschule für Wirtschafts- und
Sozialwissenschaften, Dufourstrasse 48
CH-9000 St. Gallen, Switzerland

In the last few years, the theme of self-organization has provoked a considerable
volume of research in the natural sciences, and important (but, to my mind, still
too little known) progress has been made in this field. The main question for us
was to what extent this knowledge about self-organization could be transferred to
social systems,e.g. firms. Such transfers - often leading to far-reaching conclusions - are nowadays being made, at a rapidly increasing rate, by social and management scientists, who mostly, however, have insufficient knowledge of the basic
theories. As with "social Darwinism", there is a danger that starting from misinterpretations and incorrect analogies, untenable theories will be developed relating to the social sciences. If this danger is to be avoided, a more thorough
understanding of the basic theories and research results concerning these phenomena is required than can be obtained by reading secondary literature and a few
popular-scientific essays. Basically, the question is: what are the arguments for
and against a connection between the management of social systems and knowledge
in the natural sciences, and in what areas can the natural sciences contribute
something?

Fundamental considerations concerning the taking into account of knowledge from
the natural sciences have already been expounded by H. Ulrich in this volume
(cf. the diagram of the three levels and of the systemic perspective). However,
as a result of our discussions between natural scientists and social scientists,
one must conclude that the question as to whether management science can profi-
tably use knowledge from the natural sciences cannot be answered with a straight
"yes" or "no". It appears that a differentiated position is unavoidable, so that
the answer must be "double-barrelled": on the one hand, a whole series of plausible arguments exists for a link between the two groups of sciences as far as research
strategy is concerned; on the other hand, this statement should not mislead one
into thinking that the natural sciences can make a contribution to all research
areas of systems-oriented management science, since certain aspects of the management of goal-oriented social systems are neither addressed directly nor taken into
account indirectly by the natural sciences (cf. Dachler).

The usefulness of a more profound discussion with the natural sciences (especially
biology) in the context of systems-oriented management science is suggested by the
following: the *basic perspectives* of both sciences contain the fundamental idea
that a systems approach provides a fruitful starting point and that multiple viewpoints are essential to a proper understanding of the subject matter. Biological
knowledge seems especially useful for a further development of the functional
viewpoint as used in systems-oriented management science. Both sciences are concerned with *similar questions*, and the supplementation of the leadership approach
in systems-oriented management science by a self-organisation approach offers
interesting perspectives. The subject matters of both sciences show similarities
with regard to their abstract system properties (complexity, dynamics, wholeness,
etc.).

The above is a plausible argument in favour of our discussions, but it provides no guarantee that the hopes raised of obtaining knowledge concerning content and methods will be realised. It seems appropriate in the following paragraphs to mention briefly some aspects of the interdisciplinary discussions which particularly interested us and which are not mentioned explicitly or fully enough in the contribution "Evolutionary Management".

We understand management theory as an applied science, producing knowledge which will be useful for people solving real world problems. Therefore we are looking for rules of action - rules for the design, control and development of the systems to be managed. Relying on cybernetics means above all taking into account the research on the mechanisms of control in its broadest sense. We have to look at these operating principles that have produced the many forms of life, adaptation, self-organisation, self-production, self-reference, evolution. We assume that both the firm and its environment are subject to similar control and development processes or operating principles as those that are observable in natural systems. Central to this assumption is the view that a firm *is* a self-organising system which can be designed, controlled and developed through consciously planned intervention only to a limited extent.

The research results of self-organisation and evolution bring new elements into a management theory. Terms such as autonomy, closureness, decentralized units, heterarchy, fluid organization, qualitative growth, incrementalism, freedom (innovation), unjustified variations, sense-making processes, spontaneous order, etc., which were not considered important in the past, begin to play a considerable role. Many of the traditional key words or elements such as planning, centralization, organizing, motivation, etc. will have to undergo a new interpretation. Also the authors of America's bestseller in management literature on lessons from America's best-run companies emphasize the ideas of self-organization, evolutionary management or the use of spontaneous order (cf. Peters/Waterman 1983). But in my opinion these phenomena are far more complex than described therein. Self-organisation does not simply mean the creating of internal competition, and evolutionary management is not only about raising the "blind variations", as Peters/Waterman put it very simply. (1)

I assume that by pointing to cybernetic principles it will be possible to help managers in finding other and perhaps more adequate solutions for the design, control and development of purposeful social systems. This may come about by deriving rules of action or confirming or rejecting the prevailing and accepted managerial

---

(1) To give some examples, the authors write in their study (Peters/Waterman 1983): "There are fundamentally two ways to sort things out in organizations. The first is "by the rules", or by algorithm, which the rationalists would have us do. It's in the nature of bureaucracy, which is defined as rule-driven behavior, to proceed this way. Thus, we find a 223-committee structure involved in new-product sign-off. At the other end of the spectrum, the "market" is brought inside. The organization becomes driven by internal markets and internal competition." (p. 215) ... "Internal competition as a substitute for formal, rule- and committee-driven behavior permeates the excellent companies. It entails high costs of duplication-cannibalization, overlapping products, overlapping divisions, multiple development projects, lost development dollars when the sales force won't buy a marketer's fancy. Yet the benefits, though less measurable, are manifold, especially in terms of commitment, innovation, and a focus on the revenue line." (p. 218). Writing about evolution they say: "Adaptation is also too complex to manage by rules in a big enterprise, so astute managers simply make sure that enough "blind variations" (i.e., good tries, successful or not) are going on to satisfy the laws of probability-to ensure lots of bunt singles, an occasional double, and a once-a-decade home run." (p. 106)

rules of action. Thus, such management rules can have their origin in a vast body of cybernetic knowledge (cf. Probst 1981). Talking about cybernetic principles for the management of social systems, it must be recognized at the outset, however, that it is illusory to hope for all-embracing principles for the solution of managerial problems.

Any of these principles can only deal with one aspect of the whole situation, and only a network of principles or hypotheses may provide managers with a tool to solve complex problem situations. But what are these principles about? Research in this approach to the management of social systems is still very young and incomplete. Therefore let me try to give briefly an impression of how cybernetic principles or hypotheses regarding laws could be used and of what kind they are (see also Probst/Gomez 1982).

Let me first give you some background information about an actual or possible managerial situation. There is a widespread tendency in American corporations to become captured by a grand strategy, so much so that they tend to ignore the actual market place, the customers as well as the problems of implementation (cf. Peters/Waterman 1983). Contrary to the common myth that also the Japanese had a master strategy, I think there are many good reasons for my assumption that, when penetrating the high quality small car market in the Western world, the Japanese did not (only) start with such a strategy. They rather manufactured what they were accustomed to making in Japan and tried to sell it abroad. This does not mean, on the other hand, that there was no strategic thinking at all concerning growth, quality, market shares and exporting to foreign countries. But what rather saved "Japan's Toyotas, Datsuns and Mazdas from near failure was the cumulative impact of "little brains" in the form of salesman and dealers and production workers all contributing incrementally to the quality and market position these companies enjoy today. Middle and upper management saw their primary task as guiding and orchestrating this input from below rather than steering the organization solely from above on a predetermined strategic course.

A cybernetic interpretation of this phenomenon could start with the following statement by Hayek (1967, p. 24): "There is no reason why a polycentric (spontaneous) order in which each element is guided only by rules and receives no orders from a center should not be capable of bringing about as complex and apparently as "purposive" an adaptation to circumstances as could be produced in a system where a part is set aside to perform such an order on an analogue model before it is put into execution by the larger structure." And the fact is rather that we can preserve an order of such complexity only if we control it *not* by methods on central and detailed "planning", i.e. by direct orders, but, on the contrary, by formation of a spontaneous order based on general rules. This also reminds us of Warren McCulloch's famous principle of redundancy of potential command where information constitutes authority. The structure is heterarchical rather than hierarchical, the command structure is completely distributed (McCulloch 1970, von Foerster 1981).

This leads to the conclusion that most often systems are not what they were planned to be. Thus, dealing with complex social systems, we do better to follow Hayek's idea "that social systems often are the result of human action but not of human design." A consequence would be the getting of more information about the sensitivity of a system (how a system might behave) and the inclusion in an iterative way of the information about the main variables of behaviour. The problem with central and detailed planning activities lies in the fact that a planner never has enough information about the system, whereas through the emergence of spontaneous orders any part of the system makes use of its available information, and the interaction of the parts leads to the desired regularities. An important cybernetic principle in this context might be the following that is based on Hayek's research.

Spontaneous orders are more complex or can become more complex than planned, man-
made orders. The more complex the system, the more we are dependent on its self-
organizing forces.

There are several consequences that could be derived from such a principle for
the management of social systems. To be more explicit concerning the above-mentioned example, planning within the framework of this principle means detecting
the rules which govern the self-organizing forces of the system. And it looks as
if incrementalism, "muddling through", planning in small steps may prove to be as
important as designing grand strategies. Depending on the given context one way
or the other is more appropriate or promising and the different ways should rather
be seen in a kind of interplay.

Other phenomena could be similarly explained in my opinion by relying,i.e.,on
the idea of autopoiesis (cf. Varela 1979, Maturana/Varela 1975). Living (or viable)
systems are autopoietic, they produce and regenerate themselves (cf. Maturana
1980, 29).

They maintain their identity. Thus the constitutive feature of autopoietic systems
is that they maintain their own organization homeostatically invariant under condition of structural change. In the concept of autopoiesis, organization is characterized as the relations that make a system (a composite unity) a unity, the
structure is defined as the actual relationships which exist beween the components of the system. Whenever the organization of a system as a whole changes,
the system itself changes and will form a new whole with a different identity;
whenever the structure changes but the organization is maintained, the system
keeps its identity. A new structure may well be necessary to cope with a changing
environment. Change of organization in a company would,e.g.,mean that we drop one
of the main parts or mechanisms that make a system what it is; that means it loses
its identity. A good example might be a company,e.g.,that disposes of its sales
department. By doing so, it changes its identity, as it is no longer in touch with
the market.

I think that knowledge of the phenomenon of autopoiesis and studying the following
principle could help us to understand much better the behaviour of institutions.

*It is a constitutive feature of an autopoietic (and therefore viable) system that
it maintains its own organization homeostatically invariant under conditions of
structural change.*

Let me try to give examples of recommendations for managers that we could possibly
derive from this principle. If the behaviour of a social system does not seem to be
purposeful to the observer, he should try to interpret it from the point of view
of the system's tendency to maintain its identity. But this principle might also
be relevant for the design of a social system. So restructuring of a social system
should always be governed by the idea of maintaining the basic organization that
has proven successful in its environment. A very important idea in this context
might be to ask whether the system we look at *is* and *should* be viable and thus
autopoietic, a question never asked in treating biological systems (see the contribution of Peter Dachler in this volume). Thus even accepting that principles
of autopoiesis do include constraints on their being extended to social systems,
or rather that they do not allows us to answer all the questions that social
systems imply, they contain two aspects:

a) they are describing mechanisms that could help us to understand
   what happens in social systems, and

b) they require a complementary view (observing systems) and
   understanding of the management of social systems.

The first point has been described before (Probst 1981, Probst/Gomez 1982, Beer 1981), the second has occupied Peter Hejl (1982), von Foerster (1981), N. Luhmann (1983) extensively and asks in my opinion for a complementary view to contingency approaches in mangement by an organizational closure-type view. Social systems exhibit an internal determination or self-assertion (as biological systems do). This view also gives us an epistemological foundation for the idea stressed in the paper on "Evolutionary Management". The manager is part of the managerial system; he is introducing himself   to the problems of organizational design and control. The logical structure thereof, where a system is included, is described by the organizational closure-type thinking. As Heinz von Foerster would say: at the moment when you have this type of closure you reduce the degrees of freedom of argumentation because the beginning of your argumentation must fall to the end. That means the whole system must be a closed system and not any solution is possible. These are self-referential problems (see the contributions of Heinz von Foerster and Peter Hejl in this volume).

References

Beer, S. (1981)   Brain of the Firm, Chichester

von Foerster, H. (1981)   Observing Systems, Intersystems Publications, Seaside Cal.

Hejl, P. (1982)   Sozialwissenschaft als Theorie selbstreferentieller Systeme, Frankfurt, New York

von Hayek, F.A. (1967)   Studies in Philosophy, Politics and Economics, Chicago

Luhmann, N. (1981)   Politische Theorie im Wohlfahrtswandel, München, Wien

Maturana, H.R. (1980)   Man and Society, in: F. Benseler/P.M. Hejl/W.K. Köck (Eds.), Autopoiesis, Communication and Society, The Theory of Autopoietic Systems in the Social Sciences, Frankfurt, New York

Maturana, H.R./   Autopoietic Systems, A Characterization of the Living
Varela, F.J. (1975)   Organization, Biol. Comp. Lab. Rep. 9.4, Dept. of Electr. Engin., Univ. of Illinois, Urb. III; Reprinted in: id. 1979, Autopoiesis and Cognition, Boston Studies in the Philosophy of Sciences, Boston

McCulloch, W.S. (1970)   Embodiments of Mind, Cambridge Mass.

Peters, T.J./   In Search of Excellence, New York
Watermann, R.H. (1983)

Probst, G.J.B. (1981)   Kybernetische Gesetzeshypothesen als Basis für Gestaltungs- und Lenkungsregeln im Management, Bern, Stuttgart

Probst, G.J.B./   New ("Second-order") Cybernetics against Mismanagement,
Gomez, P. (1982)   in: R. Trappl (Ed.), Cybernetics and Systems Research, New York, pp. 437

Varela, F.J. (1979)   Principles of Biological Autonomy, New York, Oxford

Some Explanatory Boundaries of Organismic Analogies for the Understanding of Social Systems

P. Dachler

Hochschule für Wirtschafts- und Sozialwissenschaften, Dufourstrasse 50
CH-9000 St. Gallen, Switzerland

A long tradition in the analysis of social institutions incorporates organismic analogies and metaphors as basic components into general conceptual frameworks which serve as the general conceptual context within which more specific theoretical propositions that can be empirically tested are derived. A part of this tradition involves the attempt to extend biocybernetic and general systems principles whose origin comes from research on natural systems, to the analysis of complex social institutions. To the extent that these general perspectives of social systems employ metaphors and analogies from natural systems, it is important to investigate the explanatory or theoretical benefit that they provide for a better understanding of social institutions.

The question about the heuristic efficiency and effectiveness of metaphors and analogies is not one usually raised in connection with the search for criteria by which the adequacy of scientific theories is or should be evaluated. In fact, the easy use of analogies and metaphors as a way of gaining new insights about complex social phenomena from better understood phenomena or easier to deal with models has often been criticized as a not fully legitimate scientific approach to theory building (cf. Rapoport, 1968; von Bertalanffy, 1968).

Nevertheless there have been compelling arguments in the literature, particularly with respect to the social sciences, to the effect that metaphors and analogies by necessity play an integral part in any theoretical perspective or general meta-theoretical framework (cf. Black, 1961; Schön, 1963; Hesse, 1966, 1976; Barbour, 1974; Brown, 1976). Thus the question about the heuristic adequacy of metaphors and analogies may refer to an important criterion in evaluating the explanatory potential of basic theoretical frameworks concerning social systems (Keeley, 1980). Overall perspectives or paradigms about social systems are in any case nearly impossible to evaluate on the basis of traditional positivistic criteria of scientific validity (cf. Gergen, 1982). It is therefore understandable that the great debates over different perspectives of organizational phenomena are carried out more on the basis of the adequacy of implicitly or explicitly stated analogies and metaphors used in the conceptual frameworks under discussion than on the basis of traditionally generated empirical evidence in the positivistic tradition.

While various criteria for the evaluation of metaphors and analogies have been proposed and discussed in the literature (cf. Black, 1962; Hesse, 1966; Brown, 1976), in one way or another they depend upon the extent to which the metaphor or analogy and the phenomenon to which they are applied share certain fundamental *characteristics* as well as share similar *processes* by which these characteristics became interconnected. If a metaphor analogy from one domain cannot capture the essential nature of the domain to which it is applied, then its surplus meaning for the new domain or its conceptual reframing function cannot be effective in providing a more appropriate perspective of the domain under study. In this case no more is accomplished than an act of relabeling. This may give the impression of having provided a new explanation, but in reality does not in itself open new vistas for action and may in fact be misleading by providing a false sense of having gained practical insights.

Clearly, there is, and in fact has to be, a correspondence in functioning between natural and social systems. This correspondence has allowed useful extensions of natural system principles for a better understanding in particular of the complexity of social systems. However, the metaphoric effectiveness of organismic metaphors and analogies is significantly reduced with respect to two types of fundamental differences between natural and social systems, namely the *nature of the elements* that constitute social systems and the nature of the *relationships among elements* from which form, purpose, and development of social systems emerge.

This paper therefore attempts to outline some of those fundamental characteristics of social systems whose basic nature seems difficult or impossible to capture by organismic analogies and metaphors. We will then have a better basis to look at the heuristic value of biocybernetic and general system principles for developing a more adequate understanding of complex social systems. At this juncture it may also be well to remember that the more the traditional social science perspectives or paradigms move away from their fundamental mechanistic stimulus-response perspective, away from treating fundamental change as error in the drive to develop space-time independent basic laws of nature, and away from the positivistic idea of individual-oriented dualism regarding the subjective nature of people and the objective nature of the world around them, the more the difference in *meaning* between natural and social systems may reduce the effectiveness in extending the principles of natural systems to the understanding of complex social systems.

Fundamental properties of social systems and their human elements

In describing the fundamental properties of social systems one has to recognize a basic dialectic between two highly interdependent but conceptually distinguishable levels of analyses, namely the characteristics of the individual human elements that constitute social institutions and the properties of collectivities or social systems seen as a whole. The social sciences have long struggled with this dialectic. While most theories of social systems have focused primarily on the individual level of analysis, and as a result see collectivities merely as the (usually linear) aggregation of individual properties, some systems approaches, certainly starting in a formal manner with Max Weber's analysis of bureaucracies, have attempted to view collectivities as totalities whose properties are in some sense more than the sum of the properties of their elements. Since the use of organismic metaphors and analogies are often also applied either to the individual or the collective level of analysis, it becomes necessary to describe separately the fundamental properties of social systems at the two levels of analysis. An important question concerns the ability of bio-cybernetically oriented analogies and metaphors to shed new light on the interrelationships between the individual and collective level of analysis.

Properties of social systems at the individual level

The conscious and reflective-interpretative nature of individuals.

The fundamental property of social systems which separates them in an absolute way from other systems is the conscious and reflective-interpretative nature of the human elements that constitute social systems. Although in the analysis of systems one often distinguishes between three levels, the physical, the biological and the subjectively conscious-interpretative level (cf. Ulrich, 1983), only social systems can be dealt with on all three levels. Natural systems, with the exception of humans seen as natural systems, do not contain a subjective awareness and reflective consciousness level. And even with humans seen as natural systems, the reflective consciousness dimension is a meaningful analysis dimension only if the person is viewed as a complex *totality*, since the elements themselves composing the person as a natural system do not show any self-awareness. It is this fundamental difference between natural and social systems which probably sets absolute limits to the explanatory usefulness of organismic metaphors and analogies for providing new insights about social systems. In any case, the fact that organismic metaphors and analogies are used at all as meaningful conceptual tools for the analysis of social phenomena is due either to the view that social systems are constituted by human elements which are seen as biological systems, or due to the perspective that social systems as a whole function similarly to living, goal-oriented, adaptable and evolving organisms, as if social systems behaved similarly to a "super human being" (cf. Beer, 1972, 1979; Kimberly et al., 1980). Both of these conceptual reasons for the meaningful use of organismic analogies and metaphors in the analysis of social systems become highly questionable in the context of the reflective-interpretative consciousness that characterizes the elements of social systems.

Given this state of affairs, it becomes necessary briefly to discuss the basic characteristics of the reflective-interpretative consciousness at the individual level in order to show later in what way this unique dimension of social systems plays a role in the collective processes of complex social systems. In view of the enormous literature that has dealt extensively with these problems, primarily at the individual level, only the most critical aspects of human consciousness for the evaluation of the adequacy of organismic metaphors and analogies for the study of social systems can be highlighted in the space available here.

The cyclical nature of consciousness.

One of the most difficult problems in the social sciences concerns the reciprocal interdependence between subjective human consciousness, which seems to follow "symbolic rules and processes", and the "objective" reality in which people act. The latter reality is a material world which seems to follow objective, time-space dependent laws as well as contains historical-cultural events which again do not seem to follow what we have come to accept as natural laws.

The modern discipline of psychology in particular has attempted to understand human consciousness and information processing primarily on the basis of natural laws dependent and linear computer perspectives. However, recent theory and research derived from sources as various as epistemologists like Jean Piaget (Piaget, 1952, 1954), and a rather heterogeneous group of social scientists like Bateson (1979), Neisser (1976), Segal (1971 a, b), as well as various philosophers who have worked out similar concepts, has maintained that human perception and insight is fundamentally a proactive and reality constructing interpretative process which works according to complex circular or spiral chains of determination. This view stands in fundamental contrast to the reigning concept of the passive processing of objectively available and mechanically screened information in the input-black-box-output tradition of systems thinking (Neisser, 1976).

Bateson (1979, p. 21) very clearly states the questions that are raised here:

".. the immediate task ... is to construct a *picture* (analogy) of how the world is joined together in its *mental aspects*. How do ideas, information, steps of logical or pragmatic consistency, and the like fit together? How is logic, the classical procedure for making chains of ideas, *related to an outside world* of things and creatures, parts and wholes? Do ideas really occur in chains, or is *this linear ... structure imposed on by scholars and and philosophers? How is the world of logic, which eschews "circular argu-*

*ment" related to a world in which circular trains of causation are the rule
rather than the exception?* (Underlining and comments by the present author.)"

Neisser's (1976, p. 20 ff.; p. 110 ff.) conception of the perceptual and reality
construction process is perhaps a useful illustration of the circular nature of
consciousness.

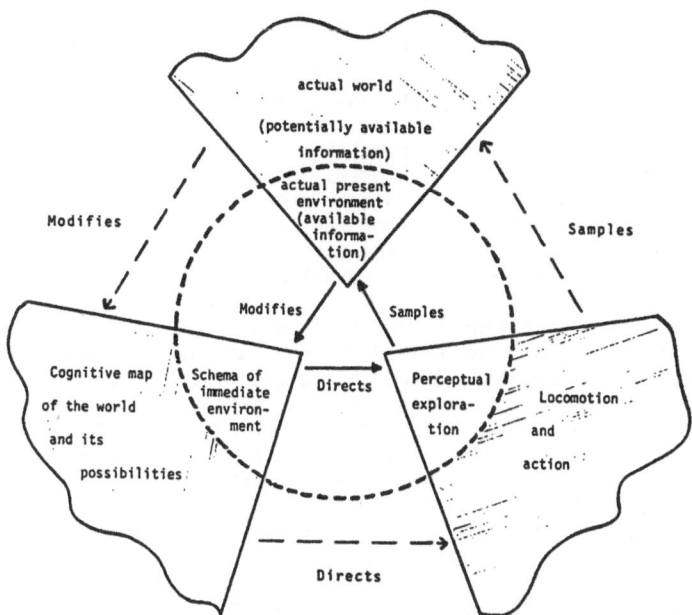

<u>Fig. 1</u>: Perceptual cycle embedded in actively constructed,
more inclusive, hierarchically or spirally ordered
cycles of exploration and information construction
(Neisser, 1976, p. 112).

Basically Neisser draws attention to the fact that understanding and perception,
i.e., the fundamental process governing the interaction of the person and his/her
environment, develops through the interplay of three crucial aspects of consciousness; namely, the preconception of that which is to be perceived and understood,
the perceptual and behavioral exploration of the world, and that aspect of the
equivocal potential information available in the environment that is selected for
further consideration (see also Weick, 1979).

Following the principle that ".:. we can see only what we *know how* to look for
(Neisser, 1976, p. 20)", perception must be based upon an already existing preconception of what is to be seen or understood. Without some existing cognitive
structures nothing can be perceived, just like cognitive structures cannot be developed or cannot be changed without perception (see also Piaget, 1952, 1954).
Neisser calls these preconceptions "orienting schemas" in order to call attention
to the fact that these preconceptions orient the perceiver within the complexity
and equivocality of the immediately available environment by constructing certain
expectations about relevant information which in turn allow the perceiver to
accept certain information out of the "buzzing confusion" in the environment as
meaningful or sensible.

With respect to the second aspect in Neisser's perceptual circle, Bateson's (1979, p. 104 ff.) observations that perception and understanding are not based on stationary constants, but rather on relationships of differences, are highly relevant. There must be at least two aspects whose relationship or interaction are perceived as a difference, a difference which has to be sufficiently large in order to overcome physiological thresholds as well as thresholds contained in the cognitive structures. "The unchanging is imperceptible unless we are willing to move relative to it (Bateson, 1979, p. 107)". Neisser similarly argues that locomotion and exploration are a necessary precondition for the recognition of changes and differences that constitute information. Perceptual exploration and locomotion, however, are not based on *random* trial and error processes (although trial and error are clearly also implied here), but are directed in good part by the expectations contained in the schemas. Orienting schemas contain in a certain sense plans about that which ought to be perceptually explored. These "plans" result in a certain readiness to "see" certain information, whereas other potential information is not recognized. As a result a selection from the total complexity and equivocality of potential information available in the environment is made, which brings us to the third aspect of Neisser's perceptual circle.

What is *selected* on the basis of the orienting schema and the perceptual exploration and locomotion represents the actually perceived environment. But this perceived environment is a "constructed" reality, since it represents only one selected aspect of the total existing complexity and equivocality. Given different orienting schemas and perceptual explorations directed at other aspects of the environment, a different reality might emerge. It is the selectively perceived and understood information from the environment which modifies the cognitive schema regarding the immediate situations of concern. The modified schema in turn directs further perceptual exploration based on somewhat changed expectations, which leads to a change in the kind of information that is perceived in the environment as well as a change in the interpretation (the assigned meaning) of the selected information, and this somewhat changed reality again modifies the cognitive schema, and so forth.

It is clear from Neisser's perspective about perceptual processes and insight, that awareness and consciousness cannot be understood as linear stimulus-response or input-output causal chains. As a matter of fact the perceptual process cannot be meaningfully separated into independent and dependent variables. Consciousness and understanding emerge out of interlinked processes as a whole.

A final and crucially important issue which is illustrated by Neisser's perceptual cycle is the embeddedness of any momentary perceptual process in the context of higher level, more abstract perceptual cycles. Thus the cognitive schema or preconception regarding what is momentarily to seen and known is embedded in higher order mental maps. Just like specific rules of conduct get their meaning from a higher order category regarding the nature of the social situation in which some interaction takes place, preconceptions of a particular aspect of the environment also get their meaning from more abstract cognitive maps which include information or knowledge about value positions and relations among objects, events, and ideas relevant to the particular perceptual event. But contrary to traditional assumptions regarding successive phases of perception from the specific to the abstract, the cognitive structures of different degrees of abstraction are simultaneously active, providing orienting patterns for the *activity* of perception and exploration at different levels of abstractions.

Similarly, perceptual exploration of some specific environmental aspect is also embedded in higher order rules of locomotion and action sequences or action patterns. Knowledge and what is selected for further thought is crucially dependent on action. But it is not only a specific (exploratory) action through which aspects

of the environment reveal themselves, it is also the more abstract behavior patterns and movement sequences which select the macro aspects of the world around us,
which in turn modify the cognitive structures at the various levels of abstraction.
While space will not allow a more specific treatment of the crucial mechanisms of
action in the process of perception and understanding, it is important to remember
that such action also implies the human ability for thought experiment, i.e., the
capacity to imagine different action sequences and their possible consequences in
different contexts. In addition, the art of asking questions - which is clearly
directed by the cognitive schemas and maps, but these cognitive structures have
to be translated into linguistic form, a process which is cumbersome due to the
fact that language can never "paint" all aspects of the "picture" or theory contained in the cognitive structure - also is a form of action which is strongly involved in the reality construction process. The way we ask questions and the kind
of questions that are asked have a great deal to do with the kind of answers we
are able to perceive. On the basis of the fact that selected information (i.e., answers we get) modifies the cognitive structures, the linguistic and psychological
processes involved in the behavior of asking questions is an often overlooked
corrective to cognitive structures. In this sense the important relationship between cognition and behavior is not the traditionally assumed one-way causal effect
from cognitive contents to behavior (as most of the decision theories, for example,
seem to imply), but a complex reciprocal interaction between the guiding function
of cognitions for action, through which, on the basis of the information made available *through action*, cognitions are in turn modified, and so forth. What Neisser's
perceptual cycle actually points to is an unending spiral of constantly changing
perception and interpretation processes through which the perceiver participates
in constructing constantly changing realities (cf. Weick, 1979).

It is important to notice that the selective processing of environmental information, which is directed by the cognitive structures and by action sequences that
include a certain amount of chance and trial and entire processes, constructs a
certain portion of the entire potentially available reality by partly reducing the
complexity and equivocality of the world. At the same time these selected realities
which are given meaning within the higher order perceptual cycles also modify cognitive structures at the various levels of abstraction. Thus the reality construction process is not merely a fiction of the imagination, but it is tied to reality,
albeit always to a selected aspect of reality. However, the momentary constructed
reality, which serves to reduce the enormous complexity and variety in the environment, can always be (and usually is) changed through the bringing to bear of a
different set of cognitive schemas and mental maps, i.e., different orienting patterns, thus resulting in a different kind of reduced complexity and equivocality.
Reduced complexity is therefore always only one posibility and therefore temporary.
The whole complexity is potentially always in existence (cf. Luhman, 1971; Bateson,
1972; Weick, 1979; Wilke, 1982).

The central role of meaning in cognitive functioning.

Another crucial property of awareness and knowing at the individual level of social systems is what Frankl (1979) has referred to as the human search for meaning
or what Weick (1979) has referred to as sense-making processes at the collective
level. From various camps of human inquiry, the humanities, philosophy and epistemology, as well as recent research and theory in the social sciences, has come a
general recognition that the nature of man as well as the nature of organizing in
our complex societies is fundamentally meaning-based as well as an evolver of meaning (cf. Luhmann, 1971, p. 30; Frankl, 1979, p. 155 f.; Pondy/Mitroff, 1979;
Weick, 1979, p. 91 f.). While there is a large literature on the meaning of meaning
and up to now little consensus seems to have been reached on the theoretical
meaning of this construct and in what way it can be used in better understanding

complex social systems, it is clear that this so basic aspect of human affairs has
been largely ignored in our attempts to understand better the nature of man as
well as the nature of collective functioning.

A very brief sketch of the role of meaning in cognitive processes would have to
include at least the following basic issues. Human consciousness is a symbolic-
interpretative process of reality construction based upon a fundamental inclina-
tion or necessity which develops out of an inherently equivocal world and out of
a basic ambiguity about one's own role in this world, to experience a meaningful
or sensible reality. In many respects this is basically a theoretical or cognitive
process of perceiving and knowing and therefore is principally a linguistic matter.
Through the capacity of individuals to respond to the interpreted meaning or to
the conceptualization of some stimulus, rather than to the "objective" properties
of the stimulus itself, individuals are essentially free from causally direct
stimulus control (cf. Dulany, 1968; Gergen, 1982). Furthermore, since the linguistic process allows the symbolic interpretation of the present by bringing to bear
a constructed reality of the past as well as realities imagined for the future,
one has to recognize the perennial gap between experience and language, i.e.,
the enormous variety of interpretations or meanings that can be assigned to experience. Following Gergen (1982), it is not a spatio-temporal world independent
of the theoretical language with which the perceiving and knowing person is concerned. His theories are not underdetermined by observation, as traditional epistemology would have us believe; observations as such cannot by themselves serve
as empirical anchors or absolute truth criteria to most descriptions of social
phenomena, because it is not spatio-temporal events to which the linguistic representations and perceptual-cognitive processes refer, but to their selected and
constructed meaning. In this sense, human understanding based upon the processes
we have very briefly outlined is not a matter of finding truth through a *direct*
correspondence to an independently (from the observer) existing objective reality.
Instead, the meaning of actions, events and situations is a *negotiated* outcome
(Weick, 1979). Cognitive interpretation of reality cannot be evaluated outside the
social context or outside the organizing processes within which social actors are
embedded. Thus sense-making is a social product. We therefore have to turn to the
collective levels of social systems and some of their basic characteristics, in
order to understand better the way in which the basic properties and processes described at the individual level may find their reflection at the collective level.

Organizing as a process of sense-making

As briefly pointed out earlier, social science thinking since the beginning of
systematic inquiry among ancient cultures has struggled with the basic problem of
whether social collectivities are merely the sum of their individual members or
whether the collective nature has properties of its own which are in principle not
reducible to the properties of single or individual groupings (cf. Bateson, 1972,
1979; Watzlawick, 1978; Smith, 1977, 1982; Hayek, 1967, 1972). In general this
question has been primarily discussed from an individual perspective so that an
analysis of collective processes must by necessity be less developed and less satisfactory from a theoretical point of view than is the case at the individual
level.

Interpersonal and intragroup relationships.

For ease of discussion I would like to discuss briefly the properties of social
collectivities with respect to two different but closely interconnected kinds of
relations between elements, namely interpersonal and intragroup relations as well
as intergroup relations and relational properties at the social system level as a
whole. The basic unit of analysis for collective phenomena cannot be the individual

element but the relationships among the elements and their different kinds of groupings. While this is not a starting assertion, the perspective taken regarding the nature of such relationships is of fundamental importance. Traditional social science, following the natural science model, thinks of relationships among people as the causal consequences of the properties which individual actors exhibit. Similar to the natural science paradigm in which the properties of matter determine causal relations based upon the laws of mechanics and thermodynamics, traditional social science assumes that interpersonal relations, for example, are best explained by looking at the properties of one actor that serve as the stimuli which determine, together with the properties of the second actor, his/her reaction, which in turn serves as the causal stimulus for the first actor, and so forth.

A very different perspective regarding the nature of social relations comes from theoreticians like Weick, Bateson, Watzlawick and Luhmann, who argue that relations among individuals of different groupings do not follow the laws of matter but "linguistic-symbolic" laws. Following Weick's (1979) concept of the "double interact" one could think of the relations among individuals as the interlocking of the perceptual cycles among the participating actors. The cognitive structures of both actors in an interpersonal relationship contain preconditions, expectations and emotional-ideological predispositions about the *nature* of the opposite partner. In line with Neisser's perceptual cycle, such cognitive schemata (including the emotional contents) about the nature of the opposite partner are embedded in higher order mental maps regarding the nature of the context in which the interpersonal interaction takes place. The orienting function of these cognitive structures leads to focused perceptual exploration and to a selection from the total available information by both parties in the interaction. Again, it has to be remembered that the cognitive structures of both actors in the relationship concern not only the present situation, but also contain conceptions about the past and the possibilities of the future. This construction of the present reality by both actors, through bringing to bear conceptions of the past as well as the future which in the interaction process can be affirmed or radically changed, is a prime example for the impossibility of explaining social processes on the basis of causal relationships within the framework of material laws. Only rather trivial human interactions involve a mechanistically oriented interchange of physical forces (e.g., a fight). In all non-trivial and complex interpersonal relations the main ingredient of the interaction *process* is information, information selectively generated, as well illustrated by Neisser's perceptual cycle. And just as in the case of the individual perceiver, the focus of any process of organizing (i.e., forming, directing and developing the *relationships* among individuals or among various groupings) is to develop out of what is initially an equivocal and complex social-collective situation a "common" sense, a meaning, a (collective) understanding. In the collective situation this sense-making process involves the construction of what is perceived to be a sensible order (among the parts of a collectivity), which provides meaning and is essential for understanding and collective learning.

The important aspect of social relations, seen as the production and understanding of information, is that ordered meaning emerges from the existence of *multiple realities* among elements or groupings in the relational network. Each actor of one relational network may well start out, based upon his preconception of the existing reality, with a meaningful "picture" for him of the relational network. But the reality one actor creates of such a collective situation may not be the same reality some other actors have constructed, based upon their perceptual, sense-making cycles.

For example, various researchers (cf. Alderfer, 1977; Kanter, 1977; Smith, 1982, 1983) have suggested that perceptions of events, people and situations are strongly "biased" by the power position from which the perceiving is done. Within the present conceptual framework this would mean that a powerful person (however we

want to define power for this discussion) is likely to have a rather different
theory of his or her counterparts and of the context of the relational network.
This person will therefore "see" different aspects of the initially equivocal relational situations (and of course other potential information is ignored or can
not in principle be perceived or understood), than those perceived and understood
by less powerful people based upon their common-sense theories or cognitive structures.

As the relational network grows in complexity and equivocality, i.e., as different
realities of the partners are intertwined and therefore combinations of very different and often contradictory or paradoxical information are generated by and
about the different partners and the situational context, one invariably observes
organizational processes like the posturing of different actors or subgroupings
(Smith, 1982). Through posturing different perspectives, interpretational rules
and other order inducing structures are selected and provisionally imposed on the
equivocal situation, in order to reduce the complexity and equivocality of the relational network (Weick, 1979). Evolved structures such as rules of procedure or
symbolic articulation of goals all have *collective informational character*, in
that such structures begin to make statements about the nature (the meaning) of
the collectivity. Weber (1978), for example, attempted to spell out through his
defining characteristics of bureaucracies the rational (i.e.,based upon a well-
articulated perspective of legitimacy) but distant or remote control or domination
of collectivities. His insights about complex social systems stand in contrast to
the currently dominating assumptions regarding person-based control through compelling supervisory commands made by individuals for the purpose of controlling
individuals. These person-oriented control processes are seen as the primary
means of designing, directing and developing complex social systems (e.g.,managing
through people). While inter*personal* command relations clearly play a role in
social systems, their impact on the collective structures and on the sense-making
processes out of which such structures emerge must be trivial in comparison to
the collective organizing processes Weber already saw and people like Weick, Bateson, Watzlawick and others are now analyzing (although from quite a different perspective to that of Max Weber).

In the space available here only a general conceptual outline of the collective
properties at the interpersonal and intragroup level can be provided, leaving a
host of theoretical problems unanswered. The crucial point, however, to be made
for the purposes of this paper is that while collective sense-making and knowing
cycles may well be *initiated* by individuals and, based upon their preconceptions,
may be handled in a goal-oriented and in that sense rational manner, in interactive
communication they produce collective realities (meanings) which emerge from social organizing processes and cannot in principle be reduced to the conceptions
and behaviors of single actors or single, indentifiable groupings. Thus design,
control and development of collectivities does not have a monocentric origin, nor
do the components of collective meaning, such as common goals, strategies and
collectively articulated actions have their origins at some powerful center. Instead these are polycentric, complexly interconnected processes involving the
simultaneous intertwining of the perceptual sense-making cycles of many actors
throughout the relational network.

Smith (1982, p. 10) argues in this connection that

  "the question 'what reality?' demands that we be attentive to the
  myriad of 'meanings' man manufactures to summarize his modes of
  connecting himself to others around him and to the cosmos at
  large."

At another point (p. 9) he concludes:

"Hence, if we wish to understand 'realities', perhaps the most
central themes to be explored are the social processes that led
to the particular sets of meanings and symbols being defined as
acceptable."

Intergroup relations and relational processes at the systems level as a whole.

The differentiation between interpersonal and intergroup relations takes account
of the fact that elements of social systems are not only individuals but also
meaningfully aggregated individuals, such as groups, coalitions or cultures. While
obviously the general processes described earlier are also relevant for the intergroup and systems level relationships, there exist some additional complexities in
the nature of these "higher order" relational networks. The reason for using the
term "meaningfully aggregated individuals" in referring to intergroup or systems
level relationships is the fact that groups or systems are elements in larger
systems primarily on the basis of processes by which aggregations develop some
identity, i.e.,some conception about the meaning of the aggregation, a consciousness of who belongs and does not belong, which of course implies a "theory" of a
set of interdependent criteria and their justification which is held consciously
or unconsciously in common by the group. Randomly appearing people that quickly
glance at a display in a store window cannot in general be viewed as a meaningful
aggregation, since only under special circumstances could an identity develop in
such an aggregation. The main problem with meaningful aggregations of individuals
(or of already existing groups) is the fact that the boundaries of such aggregations are very difficult to define by the members of the aggregation or by outside
observers or "non-members". This is so because boundaries, based on the identity
defining processes, can in practice change quite rapidly (e.g.,formal or informal
regrouping processes of individual group members as well as explicit reorganization
efforts of work groups as a consequence of changing tasks). Secondly the group
identity and therefore the group boundary can be redrawn quite differently, based
upon the different and changing perspectives of the interaction partners and of
the common perspective which emerges from the political-social processes within
and between groups (e.g.,the inclusion of minority members into white, protestant,
male-dominated aggregations).

Alderfer (1977) defines a group as follows:

"A human group is a collective of individuals (1) who have significantly
interdependent relations with each other, (2) who perceive themselves as
a group by reliably distinguishing members from non-members, (3) whose
group identity is recognized by non-members, (4) who, as group members
acting alone or in concert, have significantly interdependent relations
with other groups, and (5) whose roles in the group are therefore a function of expectations from themselves, from other group members, and from
non-group members".

While this is an analytically rather complete definition of the group concept
based upon the insights derived from group dynamics, for "real world" aggregations
this definition is far too confining. With the exception perhaps of property (5)
in Alderfer's definition, all his other defining properties of groups are characteristics only in a very general sense (i.e.,without compelling theoretical or
empirical justification of the processes that constitute these properties) and certainly in largely varying degrees. Significant interdependent relations often occur
to a much greater extent among some group members than among others. In addition,
what is perceived to be a *"significant"* relationship is highly dependent on the
nature of the intertwining of perceptual, sense-making cycles among the participants. Thus the interpretation of the significance of relationships is for the

participants as well as for outside observers a very tricky, dynamic and fleeting endeavor. In addition, the differentiation of members from non-members is in principle a nearly impossible and never-ending task, since such a differentiation greatly depends upon the conceptions and order-imposing structures that are constructed in a group, as well as upon the ethnocentric perspectives which "group" and "non-group" members develop on the basis of the kinds of *interactions* in which they are involved (Campbell, 1975). Such considerations suggest that groups, as they are formally defined on an organization chart, for example, may be groups in terms of their position or function that is indicated by the organization chart. That such groups can be viewed as organs of a viable organic system is a myth, precisely because of the long-neglected fact that human groups develop different identities on the basis of the interconnected perceptual sense-making cycles, out of which different and ever-changing realities (orders) of the group and its environment are constructed.

A further complicating factor at the intergroup and systems level of relationships is illustrated by Bateson's (1972, p. 280 ff.) arguments, which were later taken up also by Watzlawick and his colleagues, regarding different logical types in relationships that are defined as information exchange or communication. If, through organizational processes (i.e., repeated, overlapping perceptual sense-making cycles of partners in a relational network), a sensible order has emerged and, if we understand order to be a framework of explicit or implicit rules, then we have to distinguish between rules of communication and rules of the context which constitute the meta-category from which the codes contained in the communication rules obtain their meaning. The difficulty of these two types of rules lies not only in the fact that they are of different logical type (Bateson, 1972), but also in a problem raised by Hayek (1967, p. 60 f), that the metarules which provide the meaning of the communication codes are not directly accessible to human consciousness. Although people can talk about the rules of the context, they do so within the logic of the communication rules. This mixing of logical types is one of the reasons why change of the communication rules brings about only "more of the same" (Watzlawick, 1978). Nothing changes *in principle*, since the metarules have not been changed and thus no change in meaning occurs. Since the metarules are not directly accessible to individual consciousness, basic systemic change cannot be accomplished monocentrically, but must occur on the basis of complex social-political processes, similarly to the unpredictable changes in societal *interpretations* of basic problems and their possible solutions. It is the higher order mental maps and the higher order perceptual cycles that seem to play a crucial role in such changes at the meta-level.

It is clear that the theoretical tools available for explaining relations at the systems level are still very scarce, leaving a lot of work still to be done. For the purposes of this paper, however, an attempt was made to show that a great deal of organizing, in the designing, controlling and changing of complex social systems, involves not causal connections and interactions based on material laws, but relationships based on symbolic-interpretative "laws", which deal with sense-making processes in light of inherently equivocal and complex environments. While theoretical tools for the analysis of relations at the intergroup and social system level are still vague and undeveloped, the more recent work regarding intergroup processes (Alderfer, 1977; Smith, 1977, 1982, 1983), political processes in social systems (Pettigrew, 1973), the managing of contradiction, paradoxes and dilemmas in social systems (Kilmann et al., 1983; Mitroff, 1983; McCaskey, 1982, Weick, 1979) and the role of myths, rituals, and other cultural phenomena in social systems (Barbour, 1974) can well serve as a starting point for understanding social systems as sense-making and self-organizing processes.

Conclusions

The use of organismic analogies and metaphors in the development of perspectives about complex social systems has heuristic value only to the extent that central properties and processes of living organisms can be conceptually reproduced in the analysis of social systems, leading to insights not available from other perspectives. There is no doubt that the use of organismic analogies and metaphors for gaining new insights about social systems has a long history, going back to Plato's conceptions about the nature of society (Barker, 1960). As pointed out earlier, at some level there are correspondences between biological systems and properties of social systems, if for no other reasons than the fact that the human elements in social systems are biological subsystems. And the biological properties of humans, the fact that their perceptual and cognitive functioning as well as their emotional characteristics are in part based upon their biological nature, have a bearing on or certainly set limits to the kind of relationships that are possible in social systems. On the other hand, if we view the human element in social systems as a *total* phenomenon, the cognitive-linguistic-emotional characteristcs cannot be understood as a dissectable aggregation of biological elements alone. Such "wholistic" properties emerge out of biological processes as phenomena which seem to follow different laws than those used at the biological level. Thus from the very complex and interdependent functioning of cells, nerve structures and chemical "message carriers", none of which show the reflective-interpretative properties of human consciousness, a new and for social systems absolutely crucial property of man emerges which is clearly more, or at least something in principle different than the sum of the biological elements (Sperry, 1977).

A similar argument needs to be made with respect to social systems' properties and processes. This paper has attempted to show the possibility that the design, control and development of social systems, while certainly dependent on the properties and actions of individual or aggregated elements, may in good part be an emergent outcome, a product of polycentrically initiated processes, that are not easily reducible to the specific properties and actions of certain social systems' elements. The biocybernetically oriented systems approaches (cf. Beer, 1972, 1979; Malik, 1983; Probst, 1981; Ulrich 1978, 1983) or the functionalist schools of system theory (Parsons, 1956) in one way or another view social systems analogously to living organisms. As such, social systems share organismic properties, as if they were super persons, whose properties can be directly derived from those of the human elements that constitute the social system (cf. Beer, 1972, 1979). While the heuristic value of organismic analogies and metaphors has certainly proven itself, based upon the long tradition of organismically oriented theorizing about the nature of social systems, it is also true that such perspectives have profoundly restricted the kind of questions that are asked about social systems and therefore the kind of insights and problem solutions that could potentially be generated. While the organismic approaches to systems theory of concern in this paper have for the most part avoided the temptation of attributing biological properties of individuals to social systems (which however other organismic theories do quite freely; cf. Kimberley et al., 1980), one can still observe some basic contradictions in the arguments of these theorists. Although these theorists apply from biological and ecological systems central analogies regarding evolutionary processes and principles of self-organization to social systems, their organismic perspectives also include basic hierarchically ordered control and design processes which are directed toward the overall goal of survival. Traditional organic system's variables, such as function, structure and processes, are applied to social systems, but usually without *content*. It may well be reasonable to assume that cells, organs and natural organisms have functions, structures and processes which evolved on the basis of evolutionary processes and whose content was provided by nature with respect to survival and reciprocal design and control of their natural functions. In social systems, however, it was argued that, firstly,

relations among systems elements have primarily informational-interpretative character which cannot be explained on the basis of existing biological principles
and secondly, that the context of functions, structures and processes is in good
part a result of the perceptual, sense-making processes out of which socially constructed order (realities, meanings) emerge. Contrary to the biocybernetically
oriented notions of theoreticians like Beer (1972, 1979), Malik (1983), Probst
(1981) or Ulrich (1978, 1983), the management of social systems, especially in
light of the sense-making character of design, control and development functions,
cannot be located primarily in some specified group (called managers). If at the
collective level polycentric sense-making processes and their design, control and
development consequences for social systems as well as the inherently self-organizing nature of many of these processes are taken seriously, then they must also be
accepted as basic social systems properties, not properties of managers or other
centrally located functional groups.

In summary, it is clear that organismic analogies and metaphors have opened up
perspectives of social systems not recognized on the basis of other conceptual
frameworks (such as the machine model of organizations which still plagues much
of the literature). On the other hand they systematically narrow the focus on potential organizational phenomena that may be of crucial significance, if we are
to get a better grasp of the inherent properties of human collective processes.
Since consciousness and the symbolic-interpretative nature of the human mind is not
accessible to compelling explanations at the biological level, most likely because
the mind is an emergent holistic phenomenon which is based on biological elements
but whose total nature follows rules and laws that are incompatible with biological principles, organismic analogies and metaphors of complex social systems are
likely to restrict new insight and in that sense likely to lead to misconceptions
and inadequate suggestions for solutions of today's pressing organizational problems.

References

Alderfer, C.P.,   Group and intergroup relations, in: J.R. Hackman and
                           J.L. Suttle (Eds.), Improving life at work, Santa
                           Monica, Ca.: Goodyear, 1977

Barbour, I.G.,   Myths, models and paradigms, New York: Harper and Row
                           1974

Barker, E.,   Greek political theory, (5th ed.) London: Methusen,
                           1960

Bateson, G.,   Steps to an ecology of mind, New York: Ballantine
                           Books, 1972

Bateson, G.,   Mind and nature, New York: Bantam Books, 1979

Beer, S.,   Brain of the Firm, London, 1972

Beer, S.,   The heart of enterprise, London, 1979

Black, M.,   Models and metaphors, Ithaca, N.Y.: Cornell University
                           Press, 1962

Brwon, R.H.,   Social theory as metaphor. Theory and society, 1976,
                           3, 169-197

Campbell, D.T.,   On the conflicts between biological and social
                         evolution and between psychology and moral tradition,
                         American Psychologist, 1975, 30, 1103-1126

Dulany, D.E.,   Awareness, rules and propositional control: A confron-
                         tation with S-R behavior theory, in: D. Horton &
                         T. Dixon (Eds.), Verbal behavior and general behavior
                         theory, New York: Prentice Hall, 1968

Frankl, V.,   Der Mensch vor der Frage nach dem Sinn, München, 1979

Gergen, K.J.,   Toward transformation in social knowledge, New York:
                         Springer Verlag, 1982

Hayek. F.A.,   Studies in philosophy, politics and economics,
                         London: Routledge & Kegan Paul, 1967

Hayek, F.A.,   Die Theorie komplexer Phänomene, Walder Eucken Institut,
                         Vorträge und Aufsätze, Bd. 36, Tübingen, 1972

Hesse, M.B.,   Models and analogies in science, Notre Dame, in:
                         Notre Dame Press, 1966

Hesse, M.B.,   Models versus paradigms in the natural sciences, in:
                         Lyndhurst Collins (Ed.), The use of models in the
                         social sciences, 1-15, London: Tavistock, 1976

Kanter, R.M.,   Men and Women of the corporation, New York: Basic
                         Books, 1977

Keeley, M.,   Organizational analogy: A comparison of organismic and
                         social contract models, Administrative Science Quarter-
                         ly, 1980, 25, 337-362

Kilmann, R.H./   Producing useful knowledge for organizations,
Thomas, K.W./   Prager Scientific Publications, 1983
Slevin, D.P./
Nath, R. &
Jerrell, S.L. (eds.)

Kimberley, J.R./   The organizational life cycle,
Miles, R.H. and   San Francisco: Jossey-Bass, 1980
associates

Luhmann, N.,   Sinn als Grundbegriff der Soziologie, in: J. Habermas
                         and N. Luhmann, Theorie der Gesellschaft oder Sozial-
                         technologie, Frankfurt, 1971

Malik, F.,   Zwei Arten von Managementtheorien: Konstruktion und
                         Evolution, in: Hans Siegwart und Gilbert J.B. Probst
                         (Hrsg.), Mitarbeiterführung und gesellschaftlicher
                         Wandel, Bern/Stuttgart: Paul Haupt, 1983

McCaskey, M.B.,   Managing change and ambiguity, Marshfield Mass., 1982

Mitroff, I.I.,   Stakeholders of the organizational mind,
                         San Francisco: Jossey-Bass, 1983

Neisser, U.,   Cognition and reality, San Francisco: Freeman, 1976

Parsons, T.,   Suggestions for a sociological approach to the theory of organizations - 1, Administrative Science Quarterly, 1956, 1, 63-85

Pettigrew, A.S.,   The politics of organizational decision making, London: Tavistock, 1973

Piaget, J.,   The origins of intelligence in children, New York: International Universities Press, 1952

Piaget, J.,   The construction of reality in the child, New York: Basic Books, 1954

Pondy, L.R./ Mitroff, I.I.,   Beyond open system models of organizations, in: Barry, M./Staw (Ed.), Research in organizational behavior, vol. 1, Greenwich, Conn.: JAI Press, 1979

Probst, G.J.B.,   Kybernetische Gesetzeshypothesen als Basis für Gestaltungs- und Lenkungsregeln im Management, Bern/Stuttgart: Paul Haupt, 1981

Rapoport, A.,   Foreword, in: Walter Buckley (Ed.), Modern systems research for the behavioral scientist: xii-xiii, Chicago: Aldine, 1968

Schön, D.A.,   Displacement of concepts, London: Tavistock, 1963

Segal, S.J.,   Processing of the stimulus in imagery and perception, in: S.J. Segal (Ed.), Imagery: Current cognitive approaches, New York: Academic Press, 1971a

Segal, S.J. (Ed.)   Imagery: Current cognitive approaches, New York: Academic Press, 1971b

Smith, K.K.,   An intergroup perspective on individual behavior, in: J.R. Hackman/E.E. Lawler/L.W. Porter (Eds.), Perspectives on behavior in organizations, New York: McGraw-Hill, 1977

Smith, K.K.,   Groups in conflict: Prisons in disguise, Dubuque, Iowa: Kendall/Hunt, 1982

Smith, K.K.,   Social comparison processes and dynamic conservatism in intergroup relations, in: L.L. Cummings/Barry M./Staw (Eds.), Research in organizational behavior, vol. 5, Greenwich, Conn., 1983

Sperry, R.W.,   Bridging science and values: a unifying view of mind and brain, American Psychologist, 1977, 32, 237-245

Ulrich, H.,   Unternehmungspolitik, Bern/Stuttgart: Paul Haupt, 1978

Ulrich, H.,   Management - eine unverstandene gesellschaftliche Funktion, in: Hans Siegwart und Gilbert J.B. Probst (Hrsg.), Mitarbeiterführung und gesellschaftlicher Wandel, Bern/Stuttgart: Paul Haupt, 1983

von Bertalanffy, L.,   General systems theory, New York: George Braziffer, 1968

Watzlawick, P.,   The language of change, New York: Basic Books, 1978

Weber, M.,   Economy and society, edited by Guenther Ross and Claus Witlich, Berkeley, CA: University of California Press, 1978

Weick, K.E.,   The social psychology of organizing, (2nd ed.) Reading, Addison-Wesley, 1979

Willke, H.,   Systemtheorie, Stuttgart: UTB, 1982

Insights, Promises, Doubts, and Questions Emerging from a Colloquium - A Summary

G. Probst and H. Ulrich

Institut für Betriebswirtschaft, Hochschule für Wirtschafts- und
Sozialwissenschaften, Dufourstrasse 48
CH-9000 St. Gallen, Switzerland

The core of the discussions in St. Gallen was the phenomenon of self-organization.
In the last few years this phenomenon has been studied by many natural scientists
and the discussions showed that up to now there have been few known and essential
progresses in this field. In the discussion the question kept arising how far such
scientific insights about the phenomenon of self-organization could be transferred
to social systems, such as firms. Social and management scientists realize such
transferences in an increasing number of cases and sometimes with far-reaching
conclusions but often without having a sufficient knowledge of the basic theories.
The danger of such a proceeding lies in the fact that - similar to a social darwinism - due to misinterpretation and inadmissible analogies indefensible social
scientific theories are being developed.

The discussions have - without any doubt - brought some success as to a better
understanding of the basic theories and research results of this phenomenon but
they have also generated many additional questions, doubts and hopes for results
in the near future.

From the variety of the discussion points some general themes shall be conclusi-
vely mentioned to give an impression of the insights, promises, doubts and questions that were repeatedly the centre of discussion.

General theme: *theory of knowledge* (epistemology/ontology)

. The central epistemological importance of the distinction between an independent
  and a participating observer was discussed repeatedly. The former speaks about
  something external to himself, whereas the latter ultimately speaks about himself, since he is a part of the observed system.

. Varela, bearing in mind the essential nature of this clear distinction, calls
  for every science to go over to "clean epistemological accounting".

. From the discussion about the ontological status of the supposedly perceived
  order of the world, two positions emerged: on the one hand, Riedl's "evolutionary theory of knowledge" position, which states that the order of the real
  world is given from the nature of things, and that this order is "discovered"
  through the evolutionary adaptedness of the human organs of perception. On the
  other hand, the position of von Foerster and Varela, who argue that the supposedly discovered order is *per se* an invention of the observer which is dependent on
  the point in time and *on* the observer's viewpoint. Our view of the world is only
  one of many possible ones; "natural laws" are only useful explanatory principles.
  There is no "real world" as an independent entity, but the order of the world is
  a result of an individual's own actions (ontogenesis) and of the whole history
  of mankind (phylogenesis). The concepts "real world" and "absolute reality"
  should be avoided in scientific discourse. (This is the consequence of radical
  construction as advocated by v. Glasersfeld, Watzlawick, v. Foerster or Varela).

. Varela summarises his conception of radical constructivism with the concept of
"groundlessness": human existence is "groundless", being characterised by total
interdependence and emptiness. Mankind interprets events and thus gives them a
sense, an order, which, however, is neither arbitrary nor attachable to some
fixed point of reference. Varela's position can be described as a middle way
which corresponds neither to solipsism or subjectivism, nor to objectivism or
representationism. This does not mean that a human being cannot feel some sense
of purpose; this is quite possible and even necessary. He should also be willing
to defend this sense of purpose; however, he must at the same time accept that,
in principle, this sense cannot be explained ontologically. As an ethical guideline for human actions, Varela proposes the following postulate: Search for some
kind of constant and permanent suspension of your temptation for certainty".

. Two differing approaches to *models* were noted: on the one hand the idea that
models are a representation *of* the world, with the danger that model and reality
may be confused, on the other hand the idea that models are only mental constructs which may help one to understand a given problem better. Thus, the use
of system-oriented terminology does not mean that the world is composed of systems, but only that systems concepts have shown themselves to be useful tools
with which to think about the world. Checkland therefore incorporates his "Soft
Systems Methodology" into a new paradigm of systems theory which starts from the
idea that reality is not ontologically "knowable", but possibly systematically
"understandable" (cf. Checkland).

General theme: *Philosophy of science/research methods*

. One of the most important discussion topics was the question of what can be considered a *"scientific explanation"*. In contrast to the classical criterion of
prediction, Varela proposes the criterion of "reproduction". According to this
criterion, a phenomenon can be considered scientifically explained if it is possible to develop a system of simple rules which produce phenomena that basically
(i.e. in their "fundamental underlying rules") correspond to the phenomena that
were to be explained. A simplifying model of this type does not need to simulate
the phenomenon in question in every detail; it is sufficient if it (the model)
underlies the *class* of phenomena that the phenomenon in question belongs to (cf.
the use of "explanations of the principle" and "pattern predictions" according
to Friedrich v. Hayek (cf. the writings of F.v. Hayek, F. Malik, G. Probst)).

. In the context of the philosophy of science, Varela stresses the distinction
between *"symbolic and operational descriptions"*. An observer can ascribe to a
phenomenon various properties which may be of use in describing and understanding it. For example, biological evolution can be described "symbolically" as a
directional process. This should be strictly separated and distinguished from
descriptions of the operational mechanisms which produce (and therefore explain)
a given phenomenon. Thus, evolution does not function "operationally" according
to a directional factor - it is directionless and opportunistic - even if it can
post hoc facto be symbolically described as directional. Directionallity is not an
operational characteristic of the evolutionary phenomenon itself, but is ascri-
bed to it by the observer.

. The essence of attempts of interdisciplinary research does not consist in simply
transferring knowledge or concepts from one area of knowledge to another, but
primarily in a mutual learning of one another's language and thought processes,
and in listening to one another. Thus, the barriers between disciplines can be
broken down, and the possibility created that a more all-embracing, common body
of knowledge may emerge, in which the connections and commonalities of things
will be better understood.

. In attempts at interdisciplinary research, it is necessary to focus on a *given system that is understandable for all*, in order to be able to progress from the relatively infertile level of general description to a more useful level of operational explanation.

General theme: *Biology and Self-Organization*

. In the course of the discussions, *two complementary forms of description of* biological systems were in evidence: an organism can be described, on the one hand, as an input/output system, on the other as an operationally closed system. In the first case, the relationship between the organism and its environment can be understood by an examination of the inputs and outputs. In the second case, however, the organism is perceived as an autonomous system for which possible influences from its environment signify only disturbances of its internal coherence. For Varela, the selection of the form of description is not a philosophical or ontological question, but rather one of the probable usefulness of that particular viewpoint for the investigation of the problem concerned. However, this complementarity on the level of the *description* of biological systems disappears when the aim is to *explain* given biological phenomena. For example, if one is aiming at an operational explanation of nero-physiological problems, the only suitable viewpoint is that of autonomy, with its emphasis on the internal self-organizing aspects of a biological system.

. The autonomy viewpoint is crystallized in Varela'a definition of living systems as *"autopoietic"* systems. Living systems are maintained as distinguishable units in physical space by their autopoietic organization. The latter is described by Varela in a recent interview as two mutually interacting processes. On the one hand, there is a molecule-production system that is able to establish a limit, a topological discontinuity; on the other hand, precisely this limit makes the system possible, because, without it, the system would disintegrate. Thus, we have a rather strange situation: the existence of the unit is totally dependent on the unit's inner dynamics, and conversely the dynamics of the unit are only made possible by the unit's existence. This is the orginal meaning of the concept of "autopoiesis", which, literally, means "self-production" (Varela). Central to the definition is the mutual conditioning of the two aspects, dynamic self-production and spatial limitation.

. Much time was occupied by the discussion between Riedl and Varela as to whether biological *evolution* could be viewed as an *optimisation process*: Riedl regards evolution as a process of gaining information about the environment which leads to increasingly optimum adaptation. Varela criticises the use of concepts of optimisation in connection with evolution for the following reason: optimisation always implies maximisation along a given dimension. However, evolution is a complex multi-dimensional event, which renders impossible the identification of a given reference point for the measuring of optimisation. The use of the concept of optimality may indeed be profitable for the symbolic description of evolution, but it cannot be used for an operational explanation of evolution. For Varela, evolution is a natural process of random drift within distant boundaries, which is not based operationally on direction or optimisation. This understanding of evolution is also expressed in Varela's *proscriptive understanding of selection*: natural selection only marks out the framework of what is possible, without precisely determining what "optimal fit" is (*prescriptive selection*), within these limits, it is the internal self-organization of the organism that determines what happens in the course of evolution.

. Biological evolution and *two-level selection*: despite the disagreement between Riedl and Varela over their respective understanding of evolution, there is quite close agreement in their views with regard to the significance of an internal selection phase as a preliminary to external selection. Riedl talks of an internal selection ("Betriebsselektion") which channels possible evolutionary stresses on the evolutionary significance of inner coherences in organisms ("multiple epigenesis"); this significance is greater than that of selective environmental influences on evolution.

General theme: *Physics and Self-Organization*

. *Synergetics and autopoiesis*: synergetic systems (e.g. lasers) are not autopoietic systems. They fulfill the condition of operational closure, but not that of self-produced delimitation. While organisms (e.g. cells) themselves produce the membrane that delimits them, the boundaries of a system that produces laser beams (double mirror) is a product of the research scientist.

. *The general validity of synergetics*: if the axioms of synergetics are fulfilled, then the discoveries of this science apply independently of the type of elements composing the system (electrons, molecules, human beings). This proposition of Haken's did not meet with universal acceptance.

. *Practical consequences of synergetics*: in order to induce a qualitative change in macroscopic systems, two steps are necessary:

1. the system must be destabilised
2. fluctuations must be introduced into the destabilised system.

Once a system has been destabilised, even small fluctuations can lead to a fundamental change in the system. However, the question must remain open as to which of the many small fluctuations will prevail and thus lead to a qualitative change.

General theme: *Systems Theory and Self-Organization*

. *Trivial and non-trivial machines/systems*: in the case of trivial machines/systems, the transformation of an input into an output occurs in such a way that, if we know both input and output, we can reconstruct the transformation process by trial and error. However, in the case of non-trivial machines, it is not possible to determine the transformation process unequivocally, despite an enormous expenditure of time on experimenting with different input and output magnitudes. This occurs because very simple steps in the transformation process are dependent on one another. The supposed simplicity of the internal transformation processes of non-trivial machines can mislead one into treating them as trivial.

. According to v. Foerster's *"order from noise" principle*, the introduction of "noise" (nonsense, undirected energy) into self-organizing systems can contribute to their viability (cf. v. Foerster; see also Haken's ideas on the significance of fluctuations).

. Self-organizing systems are characterised by *heterarchy*. Every part is potentially of equal importance. The current importance of each part is dependent on the relevance of the information that it can contribute to the system (McCulloch's "principle of redundancy of potential command, where information implies command").

. Von Foerster describes the consequences of insights into autopoiesis and epistemological research (constructivism) as both an aesthetic and an ethical imperative. The understanding and acceptance of autonomy (cf. epistemological processes according to v. Foerster, autopoieis and autonomy principles according to Varela) automatically involves responsibility: "If I alone decide my deeds, then I am responsible for my actions". Therefore, he formulates his "aesthetic imperative" thus: "If you want to understand, learn how to act appropriately." The "ethical imperative of v. Foerster's constructivistic systems theory says: "Always act in such a way that further possibilities arise".

General theme: *Self-Organization and management of social systems*

. Also in the case of social systems, there exists the choice between the *input-output approach* and the *autonomy approach*. The latter would steer research more towards the inner coherences and recurring constellations in social systems. An interesting research area would then be the search for certain "epigenetic land-scapes" in the events that occur in social systems. For Varela, however, the usefulness of employing certain analytical tools from the physical and biologi-cal sciences (e.g. multiple epigeneses) in a management context still remains an open question. He wonders whether, in view of the much greater complexity of social systems, there is not a "hiatus" between the natural and the social scien-ces, for he has so far not learnt of any really useful applications to social systems of tools derived from physics, i.e. none that enabled a better explana-tion of social phenomena. Haken is much more positive in respect to this que-stion; while not neglecting the difficulties, he urges the trying-out of such applications before one judges their usefulness (concerning the peculiarities of social systems and the difficulties that result therefrom for research metho-dology, see Malik, Probst, Dachler).

. In a management context, the distinction whether a manager is considered an in-tegral part of a social system or not is also important. If he is, the system can be described as "self-organizing"; if he is not, it looks as if the social system "is being organised".

. If a social system is regarded as self-organizing, then it follows, from the "principle of redundancy of potential command (see above) that every member is a potential leader, hence the "corresponsibility of every member".

. People and social systems are clearly "non-trivial machines/Systems". Nonethe-less, there exists a human tendency to trivialise them. It is the task of mana-gement science to counter these tendencies, because a trivialisation of social systems leads to "self-castration of the potentialities of a system for crea-tivity".

The aim of this brief summary is mainly to illustrate the wide range of themes addressed in the colloquium and in this book. Far too many questions remain un-answered, too many results are not proven yet and little known, many things are unusual and ask for lengthy processes of learning. This book can not be under-stood as a final report, it is just a beginning.

Index of Contributors

Boldface numbers indicate the page number of the contribution

Prof. Dr. Peter Checkland **94**

Professor of Management and director of the Department of Systems, School of Management and Organizational Sciences, University of Lancaster (GB). Research in the areas of problem-solving methods, systemoriented management, systems engineering and philosophy of science.

Many publications on systems thinking, - methodology, - techniques (e.g. "Systems Thinking - Systems Practice")

Prof. Dr. Peter Dachler **132**

Professor of Psychology at the HSG (*), member of the editoral board and consulting reviewer for various american professional journals, member of the governing body of the "Society of Organizational Behaviour", former associate professor of psychology at the University of Maryland.

Various publications on intelligence research, personal attitudes and motivation, learning theory

Prof. Dr. Heinz von Foerster **2**

Professor emeritus, former director of the Biological Computer Laboratories, University of Illinois, Urbana. Participant at the Macy-Conferences 1958. Mayor research on: homeostatic mechanisms; regulation, stability, communication; mechanisms of brain and mind; self-organization; second-order-cybernetics.

More than 90 important publications, especially on cybernetics, brain-research, self-organization, biological computers, artificial intelligence, information processes, cognition

Prof. Dr. Hermann Haken **33**

Professor of Theoretical Physics at the University of Stuttgart (D), Dr. in mathematics from University of Erlangen (D). Main research areas: Fest-Körper-Physik, quantum optics', laser theory and synergetics, which was started by him.

(*) HSG = Hochschule St. Gallen = Saint Gall Graduate School of Economics, Law, Business and Public Administration, Switzerland

Many publications on the above mentioned topics, the most important of which are: quantum-field-theory, synergetics, light (waves, photons, atoms), atom- and quantum-physics

Dr. Peter M. Hejl **60**

Locksmith apprenticeship, worked for many years as a foreman in the metal-industry. Studied political sciences and sociology at the Freie Universität Berlin (Dipl. Politologe), scientific collaborator at the Institut für Wissenschafts- und Planungstheorie, then at the Institut für Mediensoziologie/Medienpsychologie of the Forschungs- und Entwicklungszentrum für objektivierte Lehr- und Lernverfahren des Landes Nordrhein-Westfalen in Paderborn (D). Since 1974 lecturer of sociology at the University of Paderborn (GHS). Dr. in social sciences from Bielefeld University (N. Luhmann). Main research areas: autopoiesis, cognition and communication.

Publications: a book on social sciences as a theory of self-referential systems, diverse papers on autopoiesis and social sciences

PD Dr. Fredmund Malik **105,121**

Senior lecturer of management theory at the HSG, director of the Management-Zentrum, a consulting company, member of the board of directors of the Institut für Betriebswirtschaft, full-time researcher of the Swiss National Science Foundation for several years.

Various publications, especially on systems methodology, management methods, strategic management and "evolutionary management"

Dr. Gilbert Probst **105,127,148**

Senior lecturer of management theory at the HSG, head of research and member of the board of directors of the Institut für Betriebswirtschaft, member of various research groups, project coordinator of the Swiss National Science Foundation project on systems-oriented management.

Various publications on empirical research of values held by swiss managers and students, systems-oriented management and especially on cybernetic rules and evolutionary management

Prof. Dr. Rupert Riedl **42**

Studied medicine, anthropology and biology, head of various marine expeditions. Prof. of zoology at the Vienne University (A), former professor of zoology and marine-biology at the University of North Carolina. Main research areas: systematics, comparative anatomy, ecology and marine-biology, evolutionary theory, theoretical biology, evolutionary epistemology, systems sciences.

Many publications on marine-biology, order-phenomena in living systems, evolutionary theory and evolutionary epistemology

Prof.Dr.Drs.h.c. Hans Ulrich
80,148

Professor of management theory at the HSG, president of the Institut für Betriebswirtschaft and of the Management-Zentrum, Saint Gall, head of various research projects on systems-oriented management, changing values, management education in the private and public sector (health care systems).

Many publications on management- and organization theory, systems-oriented management, business policy, management philosophy and applied social sciences

Prof. Dr. Francisco Varela **25**

Professor of biology at the University of Santiago de Chile, adjunct research at the Brain Research Laboratories of the Medical Center of the New York University, Ph.D. from Harvard University. Extensive research on self-organization, autopoiesis and autonomy (partially with Humberto Maturana).

Many publications on neuro-physiology, cybernetics, theoretical biology, biological epistemology and philosophy of science

Further participants at the 1st St. Galler Forschungsgespräche on Management and Self-Organization in Social Systems at Saint Gall Graduate School of Economics, Law, Business and Public Administration, St. Gallen, Switzerland, 14 - 16th September 1983:

Dr. Michael Ben-Eli

Director and owner of Cybertec, New York, USA

Prof. Dr. Kurt Dopfer

Professor of Economics at the HSG

Prof. Jean-Pierre Dupuy

Research director at the CNRS, Paris, conferences at the Ecole poly-technique, Paris, F

Dr. Thomas Dyllick

Collaborator at the Institut für Betriebswirtschaft, lecturer of organization-theory at the HSG

PD Dr. Peter Gomez

Senior lecturer at the HSG, director of corporate planning at Distral Holding (Switzerland)

Dr. Walter Krieg

Senior lecturer of management theory at the HSG, director Strategic Planning and Corporate Development in the Amiantus-Group (Switzerland)

Prof. Dr. Bernd Schiemenz

Professor of Management and head of the Business Administration Department at the University of Marburg, D

Dr.des. Markus Semmel

Researcher at a Swiss National Science Foundation project on systems-oriented management

Prof. Dr. Emil Walter

Professor of social-psychology and applied social research at the HSG

W. Heisenberg

Gesammelte Werke – Collected Works

Wissenschaftliche Originalarbeiten – Original Scientific Papers

Herausgeber: **W. Blum, H.-P. Dürr, H. Rechenberg**

Ein Buchereignis: Acht Jahre nach dem Tod Werner Heisenbergs, einem der größten Physiker dieses Jahrhunderts, beginnen die Verlage Piper und Springer mit der Veröffentlichung seiner „Gesammelten Werke". Zunächst erscheinen:

Abteilung/Series A

Original Scientific Papers
Wissenschaftliche Originalarbeiten
(In 2 Teilen, die nur zusammen abgegeben werden.)
ISBN 3-540-13400-X Springer-Verlag

Abteilung/Series B

Scientific Review Papers, Talks, and Books
Wissenschaftliche Übersichtsartikel, Vorträge und Bücher

1984. Etwa 950 Seiten (509 Seiten in Englisch, 92 Seiten in Französisch, 10 Seiten in Holländisch)
ISBN 3-540-13020-9 Springer-Verlag

Springer-Verlag
Berlin
Heidelberg
New York
Tokyo

Abteilung/Series C

Allgemeinverständliche Schriften

Band I: Physik und Philosophie 1927–1955;
Ordnung der Wirklichkeit, Interpretation der Quantenmechanik, Atomphysik, Kausalität, Unbestimmtheitsrelationen u. a.
Ca. 480 Seiten. Leinen
ISBN 3-492-02925-6 Piper-Verlag

Band II: Physik und Philosophie 1956–1968;
Gifford-Lectures, Sprache und Wirklichkeit, Abstraktion und Vereinheitlichung, Goethes Naturbild u. a.
Ca. 480 Seiten. Leinen
ISBN 3-492-02926-4 Piper-Verlag

Springer Proceedings in Physics
Volume 1

Fluctuations and Sensitivity in Nonequilibrium Systems

Proceedings of an International Conference,
University of Texas, Austin, Texas,
March 12-16, 1984

Editors: W. Horsthemke, D. K. Kondepudi

1984. 108 figures. Approx. 270 pages.
ISBN 3-540-13736-X

Contents: Basic Theory. - Pattern Formation and Selection. - Bistable Systems. - Response to Stochastic and Periodic Forcing. - Noise and Deterministic Chaos. - Sensitivity in Nonequilibrium Systems. - Contributet Papers and Posters.

This volume contains the invited papers and a selection of contributed papers and posters presented at the Workshop on Fluctuations and Sensitivity in Nonequilibrium Systems, held in Austin, Texas, in March 1984.
The papers deal with the subject from a macroscopic phenomenological viewpoint and address questions of basic theory, pattern formation, bistable systems, response to stochastic and periodic forcing, noise and chaos, and sensitivity in nonlinear systems. The book contains review articles as well as papers reporting recent theoretical and experimental results. This volume will be of particular value to researchers and graduate students who wish to become acquainted with the field and to obtain an overview of its current state.

Springer-Verlag
Berlin
Heidelberg
NewYork
Tokyo

Wolfgang Pauli

Wissenschaftlicher Briefwechsel mit Bohr, Einstein, Heisenberg u.a. - Band I: 1919-1929
Scientific Correspondence with Bohr, Einstein, Heisenberg a.o. - Volume I: 1919-1929

Herausgeber/Editors: A. Hermann, K. v. Meyewnn, V. F. Weisskopf
Vorwort von/Preface by V. F. Weisskopf

1979. 1 Faksimile, 34 Abbildungen, 6 Tabellen.
XLVII, 577 Seiten (Briefe in Deutsch, Dänisch und Englisch). (Sources in the History of Mathematics and Physical Sciences, Volume 2)
ISBN 3-540-08962-4

Inhaltsübersicht: Das Jahr 1919: Auseinandersetzung mit der Allgemeinen Relativitätstheorie. - Das Jahr 1920: „Relativitätsartikel" und erste Arbeiten zur Atomphysik. - Das Jahr 1921: Dissertation über das Wasserstoffmolekülion. - Das Jahr 1922: Göttingen - Hamburg - Kopenhagen. - Das Jahr 1923: Anomaler Zeemaneffekt. - Das Jahr 1924: Weg zum Ausschließungsprinzip. - Das Jahr 1925: „Quantenartikel" und Göttinger Matrizenmechanik. -. Das Jahr 1926: Rotierendes Elektron und Verallgemeinerungen der Quantenmechanik. - Das Jahr 1927: Kopenhagener Interpretation und Quantenelektrodynamik. - Das Jahr 1928: Berufung nach Zürich - Schwierigkeiten in der Quantenelektrodynamik. - Das Jahr 1929: Systematischer Aufbau der Quantenfeldtheorie. - Anhang.

Wolfgang Pauli (1900-1958) war einer der bedeutendsten Physiker des 20ten Jahrhunderts. Im Jahre 1945 wurde ihm der Nobelpreis für sein Werk verliehen. Besonders bekannt sind Paulis Beiträge in der Quantentheorie, insbesondere das nach ihm benannte Ausschließungsprinzip.
Paulis Korrespondenz ist von außerordentlichem physikalischem und historischem Wert. Viele seiner Ideen und kritischen Beiträge wurden erstmals in Briefen entwickelt. Da Pauli mit allen führenden Begründern der Quantentheorie in regem Kontakt stand, vermitteln die Briefe ein lebendiges Bild der damaligen Problemstellungen und der Diskussion, die schließlich zu unseren heutigen Auffassungen geführt haben. In einem hohen Maße fungierte dabei Pauli als das unbestechliche Gewissen, das die Hypothesen und Theorien der Kollegen einer strengen Kritik untgerzog. Darüberhinaus gehen auf Pauli viele originelle schöpferische Beiträge zurück, die auf Grund des Briefwechsels nunmehr erstmals im Zusammenhang verstanden werden können. Die vorliegende Ausgabe enthält nicht nur Pauli's Briefe, sondern auch, soweit vorhanden, diejenigen seiner Korrespondenten. Ein umfangreicher Kommentar erleichtert das Verständnis der Korrespondenz.